RAND

Ensuring Personnel Readiness in the Army Reserve Components

Bruce R. Orvis, Herbert J. Shukiar,
Laurie L. McDonald, Michael G.
Mattock, M. Rebecca Kilburn,
Michael G. Shanley

Prepared for the
United States Army

Arroyo Center

This report presents results of Arroyo Center research on the personnel readiness of the Army Reserve Components (RC). RC units activated for Operations Desert Shield and Desert Storm (ODS/S) needed to draw significant numbers of personnel from other units, i.e., significant cross-leveling took place. Given both the continued reliance on the RC in wartime and the drawdown in active and reserve units, the personnel readiness shortfalls of ODS/S give rise to concerns for the future. Will the Army be able to deploy critical RC units at the required strength and timetables in future contingencies?

This project was designed to examine the extent of cross-leveling during ODS/S, the reasons for it, the likelihood of serious personnel shortfalls in future deployments, and, based on these findings, the types of policies that could enhance the RC's readiness to deal with future contingencies. The values chosen for particular analyses (e.g., the magnitudes of specific readiness-enhancement goals and related financial incentives) are intended to support that exploration, and the resulting policy recommendations are suggestive rather than absolute. Quantification of the precise costs and benefits of the recommended policies must be based on a controlled experiment.

The recommended policies are intended to supplement other ongoing or potential initiatives and incentives to enhance personnel readiness in the Army RC. These include, but are not limited to: the reduction of manageable losses; enlistment and affiliation incentives; reenlistment incentives; bonuses targeted to enhance fill rates and readiness levels in specific MOSs and units; compensation for commuting distance; and exposure to satisfactory training opportunities, effective leadership, and adequate resource levels in the soldier's unit.

The research was sponsored by the Deputy Chief of Staff for Personnel, U.S. Army, and was conducted in the Arroyo Center's Manpower and Training Program. The Arroyo Center is a federally funded research and development center sponsored by the United States Army.

CONTENTS

FIGURES

TABLES

Many of the units the U.S. Army plans to deploy in response to future contingencies are in the Reserve Components. Ideally, all such units would be manned at their wartime states of readiness. In reality, however, budgetary considerations and limits on available personnel and training seats make it infeasible to do so. As a result, part of the current mobilization plan is to *cross-level* soldiers between units to ensure that each unit has enough qualified soldiers for the required jobs. This practice was employed in Operation Desert Shield/Storm (ODS/S), where the deployment of reserve units was regarded as largely successful.

However, although cross-leveling can be a cost-effective means to help ensure unit deployability, it is not the ideal solution to reserve readiness problems. This is particularly true for units that must deploy early in a contingency, which may have little time to acquire and integrate new personnel. Moreover, the greater the reliance on cross-leveling to offset unit readiness shortfalls, the less the likelihood that units will have had peacetime individual and collective training adequate to permit cohesive performance of their wartime mission. The readiness problem may be exacerbated in future contingencies, which may require faster deployment of larger numbers of reserve units than occurred in ODS/S.

To obtain a better understanding of Army Reserve Component personnel readiness shortfalls, their implications, and their potential solutions, we sought answers to the following questions:

- To what extent does reserve personnel readiness fall short of goals for deployment, and what is the role of cross-leveling in resolving that shortfall?

- What types of strategies might be most effective to increase peacetime reserve personnel readiness and thereby reduce serious shortfalls and the related need for cross-leveling in wartime?

- What will these strategies cost, and what will be their related benefits in reduced accession and training demands?

READINESS SHORTFALLS AND CROSS-LEVELING

We acquired, validated, and analyzed records of Army RC personnel assignments and requirements at the time of the Iraqi incursion into Kuwait. We found that in the typical unit eventually activated for ODS/S, 63 percent of the required positions were filled with soldiers who had completed training and were qualified for that Duty Military Occupational Specialty (DMOSQ). The 37 percent shortfall was about equally divided as follows:

- Positions that were not filled with soldiers (11 percent).

- Positions that were filled with soldiers undergoing initial training to become qualified for their Duty MOS (13 percent).

- Positions that were filled with soldiers who were qualified in a different MOS but had to be retrained to become qualified for their DMOS (13 percent).

The subject matter experts we interviewed—persons directly involved in the mobilization of Army RC forces—indicated that there had been substantial cross-leveling during ODS/S. They reported that the goal was to fill the typical unit with a sufficient number of qualified soldiers to allow it to deploy to the mobilization station with a DMOSQ rate of 85 percent of required strength (C-1).[1] This figure was chosen to improve readiness to a wartime footing and to try to ensure that the unit would maintain at least a C-3 rating after the potential loss of deployability status for some members upon fur-

---

[1]C-3 requires a 65 percent DMOSQ level, C-1 an 85 percent DMOSQ level.

ther verification actions at the mobilization station.   Indeed, it proved necessary to undertake a small amount of additional cross-leveling at the mobilization stations, amounting to about 10 percent of all cross-leveling actions.   There was consensus that the major factor underlying cross-leveling was a shortage of DMOSQ personnel.

Our quantitative analysis of personnel records for the RC units activated during ODS/S confirmed these reports; it indicated that cross-leveling into these operating units approached 20 percent, helping to bring their 63 percent DMOSQ rate up to the 85 percent goal. Analysis of more recent records suggests that low DMOSQ rates are still very much a readiness issue.

To characterize the future ramifications of these shortfalls, we constructed and ran a model of the mobilization and cross-leveling process for three deployment scenarios: a major regional contingency (MRC) in Southwest Asia; an MRC in Northeast Asia; and a nearly simultaneous two-MRC scenario. We assumed that the units called for in these MRCs would be characterized by the current DMOSQ rates of like unit types. However, to provide the best possible case, we constructed an ideal future force structure perfectly matched to the requirements of the two-MRC scenario. To raise DMOSQ rates to the targeted levels, the model first optimized the DMOS assignments of personnel within an activated unit and then, time permitting, filled positions through cross-training and cross-leveling actions.

We modeled two instances for each scenario. The first permitted unrestricted cross-leveling for all activated units. We recognized, however, that this case might not be ideal for planning purposes for a variety of reasons.   These include possible deleterious effects on peacetime unit readiness from reliance on cross-leveling to address personnel readiness shortfalls, noted earlier; delays in mobilizing the RC such as those that occurred in ODS/S—where M-day was 20 days after C-day—that would substantially reduce preparation time for early-deploying units; and the fact that the force structure is not likely to be ideal, providing fewer of the needed units to deploy and fewer like units to feed them with cross-leveled personnel. Therefore, we modeled a more conservative instance for each scenario that did not plan for cross-leveling into units that had to deploy within 30 days of the start of conflict.

The results for the three scenarios were similar; we thus focus on the Southwest Asia MRC in the report. When cross-leveling is unrestricted for all units activated for this scenario, most of the personnel readiness shortfalls are addressed. Given that the force was perfectly structured for two MRCs and that this scenario requires deployment for only one of them, these results provide some, but only limited, reassurance. (Indeed, the analogous results for the two-MRC scenario do reveal shortfalls even with the unrestricted cross-leveling plan.) Results for the Southwest Asia scenario based on the more conservative cross-leveling plan tell a different story, however. In that case, 25 percent of the units required to deploy (and 30 percent of the MOSs) fall short of the 85 percent DMOSQ goal.

STRATEGIES TO ENHANCE PERSONNEL READINESS

As an aid in formulating strategies intended to address personnel readiness shortfalls in the Army RC, we reviewed earlier research on the causes of those shortfalls and legislative efforts to address them. The causes fall into two main categories. First, the reserves may not utilize experience gained in the Active Component (AC) as fully as they could. Because the Army's Active Component is smaller relative to its Reserve Components than is the case for the other services, currently a smaller proportion of Army reservists—well below half—have served on active duty. In addition, about one-third of those who do enter from the AC fill a DMOS different from the one in which they gained their active duty experience.

Second, high rates of personnel turnover continue to plague the Army RC. This has a very damaging effect on readiness. During FY93, the last predrawdown year for the RC, 17 percent of the reservists changed jobs; of these, only a third were DMOSQ by year's end, whereas among those who did not change jobs, more than 85 percent were DMOSQ. During that same year, another 20 percent left the RC altogether, reducing readiness and creating substantial accession and training requirements.

We thus defined four corresponding readiness enhancement strategies: increasing the Army Reserve Components' inventory of soldiers with prior active duty experience; increasing the match rate between entering soldiers' prior Active Component MOS and their Reserve

Component DMOS; decreasing job turbulence in the Army RC; and decreasing attrition.

To quantify the anticipated effects of these strategies on readiness, we constructed and ran a second model. This model was based on observed probabilities that a reservist with certain characteristics (prior service or not, DMOSQ or not, and grade level) would, after a year's time, still be in the reserves or not and, if so, whether that reservist would have been promoted or not, would have the same job or not, and would be DMOSQ or not. We tested the readiness enhancement strategies by changing the distribution of characteristics in the postdrawdown inventory (e.g., percentage of prior service accessions) or the transition probabilities (e.g., percentage attriting). The model took the various inventory characteristics as a starting point and applied the transition probabilities year by year while maintaining a constant endstrength overall and within grade.

Our base case consisted of the continuation of current policies and turnover rates. The analysis indicates that this would provide an enlisted DMOS qualification rate of just over 68 percent. The average number of (full time equivalent) years of job experience would be about one. That is, over the course of his career, the average reservist would have accumulated the same number of duty days in his current DMOS as he would have had he been on active duty for one year. The number of annual accessions for the RC would be just under 94,000. The associated annual training load deriving both from initial entry training and from retraining soldiers who change jobs would be approximately 180,000 training seats.

Although the Title XI legislation specifically calls for the ARNG to increase its proportion of prior active service personnel to 50 percent in order to improve readiness, the model indicates that even if it were possible to achieve this goal for the entire Army RC force (ARNG and USAR), it would result in only limited improvement. For example, in comparison with the expected effects of maintaining current policies, this initiative causes the DMOSQ rate to rise by less than 0.2 percentage points (less than 1 percent in relative terms). The increase in average number of years on the job is larger, but it still amounts to only 0.11 years (about 11 percent). Similarly, the effects on accession and training loads are modest. Overall, this change has limited effects because the number of accessions remains relatively constant, prior active accessions in mismatched MOSs must be traded for returning reservists qualified in those MOSs, and the high rate of personnel turnover eventually leads to attrition or job changes among many of the prior active personnel.

Similarly, improving the DMOS match rate of prospective prior active service accessions by 50 percent (from 65 to nearly 100 percent) provides only limited readiness enhancement relative to the base case. The DMOSQ rate rises by less than 5 percent, and annual accession and training loads are largely unaffected. The exceptions are that the job experience level improves (up 20 percent from the base case) and entry retraining is virtually eliminated for soldiers entering the reserve from the AC (but this amounts to only 5 percent of the total training load).

In contrast, the model suggests that reducing turnover offers considerably greater potential enhancement of RC personnel readiness. It indicates that reducing job turbulence by half would increase the DMOSQ rate by 9 percent and the job experience level by about 40 percent. Of course, there is essentially no effect on accessions. In contrast, there is a large reduction in the training load—about 20 percent—because reducing job changes correspondingly reduces MOS reclassification training loads.

Lowering attrition by half also has substantial benefits: It is estimated to increase the DMOSQ rate by about 8 percent; accession requirements are halved, and the training load is reduced by nearly 30 percent. There is little change in the average years of job experience because, given the greater retention, there simply is not room to access much of the available supply with prior military experience.

A promising possibility is to jointly reduce job turbulence and attrition—personnel *turnover*—which should be feasible using similar policies. The impact of reducing total turnover should be the sum of the effects of reducing each of the two problem types, since they lower readiness and increase accession and training requirements for different reasons. Indeed, as compared with the base case, the model suggests that reducing total turnover by half provides a very substantial improvement in the DMOSQ rate, to nearly 80 percent; this is twice the improvement resulting from reducing job turbulence or attrition alone. It also captures the job experience increase from

turbulence reduction and the reduction in accession and training requirements from the two separate interventions (45 to 50 percent).

COST AND SAVINGS OF READINESS ENHANCEMENT

The relevant literature on AC and RC recruiting suggests a marginal
cost of recruiting a non-prior-service accession into the RC of approximately $7,750. The cost for a prior-service accession was esti-
mated at $2,200.

Estimating the dollar savings from reductions in training load was
more complex. Our cost estimates cover all the variable costs of
training. For most cost elements, we used TRADOC published factors to calculate the variable cost of changing training load. Student
pay and allowances, one of two cost elements not covered by the
TRADOC Resource Factor Handbook, was computed by applying pay
and allowance factors from The Force and Organization Cost Estimating System and *The Reserve Forces Almanac* to the length of individual courses. The cost of ammunition for each course was obtained from the Army Manpower Cost System data file, maintained
by the Army Cost and Economic Analysis Center. Following similar
procedures, we derived per-soldier costs for four types of training:
Basic Training ($6,150); AIT for initial skill training ($7,750); AIT for
MOS reclassification ($10,200); and RC school training for MOS re-
classification ($4,900).

It is evident from the readiness enhancement analyses summarized
above that reducing reserve attrition is likely to lead to substantial
cost savings in the form of lower recruiting and training costs. However, we also expect that it would incur considerable outlays due to
the required payment of incentives. The relevant research is limited.
However, it indicates that increases in military compensation significantly reduce the rate of attrition. A 10 percent raise in average drill
pay is estimated to reduce attrition by 4.5 percent to 9.5 percent. We
used these figures to bound the marginal cost of compensation per
instance of reduced attrition at $13,200 to $27,750.

In addition to a reduction in attrition, we also recommend reducing
job turbulence as a way to enhance readiness. As is true for attrition
reduction, reducing job turbulence is likely to require financial incentives. We are unaware of research quantifying the size of such in-

centives. However, many reservists who change jobs are thought to do so to increase their promotion chances. Thus, a bonus that makes up the pay differential to the next grade could reduce turbulence. The number of bonuses offered would depend on the magnitude of the reduction needed and the ability to target the compensation to recipients who are likely to change jobs for promotion purposes. We estimate the marginal cost at $200 to $1,000 per instance of reduced turbulence.

Because the compensation needed to reduce attrition is much larger than that needed to reduce job turbulence, the former should also yield the latter's benefits if made contingent on staying in the same job.

COST-BENEFIT OF ALTERNATIVE STRATEGIES

As shown in Table S.1, a bonus paid to reduce job turbulence by half might be expected to improve the DMOSQ rate by about 9 percent, relative to current policies; depth of job experience would increase by close to 40 percent. An advantage of the job turbulence-reduction bonus is that the policy may not cost anything; on the contrary, we estimate that it would result in a net savings, because the cost of the turbulence-reduction bonus would be outweighed by savings in training costs.[2]

But the turbulence-only approach also has a drawback: Many units require improvements in their DMOSQ rates exceeding the 9 percent provided by turbulence reduction if they are to reach their targeted readiness levels. We can accomplish this by jointly tackling attrition and job turbulence reduction. If we provided a bonus to reduce total turnover—both attrition and job turbulence—by half (second row of the table), we could expect the DMOSQ rate to improve by nearly 17

---

[2]The USAR wishes to discourage homesteading—extended assignments to one unit— among senior personnel in order to provide a broader experience base. It should be noted that the proposed job turbulence-reduction bonus can be applied to enhance readiness in concert with this goal. First, consistent with other USAR readiness enhancement incentive programs, eligibility will likely be targeted to the more junior portion of the enlisted force, where most personnel turnover occurs. Stabilizing these personnel in their jobs is critical for building depth of job experience and initial leadership skills. Second, the policy is not intended to limit promotion to positions in other high-priority units in the same occupational specialty.

Table S.1

Estimated Readiness Benefits and Cost Savings
of Alternative Turnover-Reduction Policies
(Per 10,000 reservists)

Policy Yielding Reduction of	DMOSQ Improvement (%)	Job Experience Improvement (%)	Net Savings/(Cost)	
			Lower Savings ($M)	Upper Savings ($M)
Turbulence – 50%	9.1	38.0	3.6	4.2
Turbulence – 50%, Attrition – 50%	16.6	45.0	(13.6)	2.4
Turbulence – 50%, Attrition – 25%	14.7	56.6	(3.4)	4.2

NOTE: Personnel readiness improvements are relative to the base case. As shown in Table 5 (see p. 29), the DMOSQ rates for the first two rows in this table are 68.4 percent and 79.7 percent, respectively. Based on results from the readiness enhancement model, the DMOSQ rate for the last case is 78.4 percent. The job experience improvement is the percent increase in full-time equivalent job years.

percent, in relative terms; depth of job experience would increase by 45 percent. Although the potential benefits of such a policy appear to be substantial, they might also be very expensive.

However, a bonus large enough to reduce attrition by 25 percent should still be large enough to reduce job turbulence by 50 percent. The analysis suggests that such a policy could be quite beneficial (third row of the table). The estimated increase in the DMOSQ rate approaches 15 percent, and depth of job experience increases by 57 percent. The costs for this policy appear to be much more modest than those for a bonus intended to reduce attrition by 50 percent, and may actually result in a net savings.

Based on these results, what policy might best address the personnel readiness shortfalls identified for the Southwest Asia scenario, and what would its resulting cost or savings be? For units unable to deploy to the mobilization station at the targeted DMOSQ level, a bonus to reduce job turbulence by half could be employed when the required improvement in the DMOSQ rate is below 10 percent; the larger bonus designed to reduce attrition by 25 percent and job turbulence by 50 percent could be used when the needed improvement is 10 percent or greater. The estimated cost of this policy ranges from a net cost of $2.2 million to a net savings of $4.0 million.

Considering the promise of the turnover-reduction strategies, it would be to the Army's advantage to implement them in controlled settings that would allow their effects and costs to be systematically evaluated and the uncertainties resolved. Specifically, such an experiment would be able to settle questions relating to the exact size of the bonuses required to reduce attrition and job turbulence, the ability to target the bonuses, their possible market expansion and unit/skill channeling effects, and the overall scope of the readiness enhancement program that is economically practicable.

ACKNOWLEDGMENTS

We are grateful for the advice and support of our sponsors within the Office of the Deputy Chief of Staff for Personnel, Department of the Army. In particular, we wish to thank Lieutenant General Theodore Stroup, Deputy Chief of Staff for Personnel, his predecessor, Lieutenant General Thomas Carney, Major General Thomas Sikora, the Director of Military Personnel Management, his predecessor, Major General Frederick Vollrath, and Colonel Anthony Durso (retired), former Director of Plans, Analysis and Evaluation. We also are grateful to the members of the U.S. Army Reserve Command (USARC), the 77th Army Reserve Command (ARCOM), and the Defense Manpower Data Center (DMDC) who assisted us in the conduct of this work. We particularly want to acknowledge Jane Crotser (DMDC) and CPT Bill Smathers (77th ARCOM). At RAND, we are grateful for the assistance of our colleagues Ron Sortor, Jim Dertouzos, Tom Lippiatt, Mike Polich, and Patty Dey. We also would like to express our gratitude to Nikki Shacklett, for her helpful editing, to Jim Chiesa, who helped draft the initial and revised versions of this report, and to Fran Teague, for her skill and patience in helping to finalize it. Finally, Lieutenant Colonel Chad Stone (USARC, retired) played an invaluable role in the conduct of this research.

Chapter One

INTRODUCTION

This report presents the results of Arroyo Center research on the personnel readiness of the U.S. Army Reserve Components (RC). The RC contains many of the units the Army plans to deploy in future contingencies. Although ideally all such units would be manned at their wartime states of readiness, budgetary considerations and limits on available personnel and training seats make it infeasible to do so. Consequently, the current mobilization plan includes the *cross-leveling* of soldiers between units to ensure they have enough qualified soldiers for the required jobs. Cross-leveling was employed in Operation Desert Shield/Storm (ODS/S), where the deployment of reserve units was regarded as largely successful.[1]

The significant roles of the RC and cross-leveling are not surprising, but they give rise to concerns for the future. Reliance on the RC has been part of the planned response to major contingencies in the past, and this will continue to be the case. Put simply, it is not feasible or, from many perspectives, desirable to put all the units required to respond to major wartime contingencies in the Active Component (AC). Similarly, it is not reasonable in peacetime to expect every unit in the RC to meet all wartime readiness requirements. Rather, we expect to prioritize the RC units according to the timetable and order in which they are needed for major contingencies; peacetime resourcing and readiness levels follow accordingly. When necessary, cross-leveling additional personnel into mobilized units to fill duty

---

[1]An additional, small proportion of the deploying personnel were cross-leveled into the RC units from the Active Component and IRR; they are not considered in this analysis.

positions with critical shortages of qualified soldiers is part of this plan, provided it can be accomplished within the unit's deployment timetable. Such personnel are already qualified in the needed skills or are individuals who can quickly become so through special training programs.

Although cross-leveling can be a cost-effective means to ensure unit deployability, there is a price to be paid in peacetime unit readiness levels. To the extent we allow high-priority units to maintain lower peacetime fill rates and job qualification levels, relying on cross-leveling during wartime, their collective training and overall readiness will suffer. Furthermore, wartime cross-leveling may become more difficult, because future contingencies may require faster mobilization than Desert Shield (see Sortor, 1995). The current goal would result in deployment of almost twice as many personnel in the first sixty days as were deploying in ODS/S, in part because of the reduction in AC endstrength. The corresponding reduction in RC endstrength further exacerbates the potential for difficulties.[2]

Congress has recognized the importance of the reserve forces and has called for increased levels of RC readiness through a variety of programs.[3] One of these is to recruit increased numbers of soldiers with prior active duty experience into the Army RC, which is intended to increase reservists' experience levels. Such programs could help to enhance readiness, thereby reducing concerns about wartime reliance on RC units. Still, our analysis indicates that other fundamental issues—such as qualification for one's military job—need to be addressed. The underlying question is: How can the Army best ensure RC personnel readiness in peacetime while controlling costs and demands on the training system?

To answer this question, we examined the extent of cross-leveling during ODS/S, the reasons for it, the implications of these findings for the RC's readiness to deal with future contingencies, and methods of enhancing that readiness. The research proceeded in several steps (see Figure 1). We began with a qualitative analysis in which

---

[2]The planned reductions in endstrength after FY89 are from 770,000 to 495,000 for the AC and from 776,000 to 575,000 for the RC.

[3]U.S. Government, Army National Guard Combat Readiness Reform Act of 1992 (Title XI), P.L. 102-484, Sec. 1111–1137, 1992.

RAND*MR659-1*

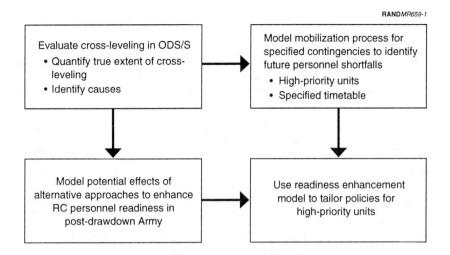

Figure 1—Analytic Approach

we obtained reports of the cross-leveling actions conducted during Operation Desert Shield/Storm from subject matter experts (SMEs) involved in the mobilization and deployment effort. We were particularly interested in their reports of the extent of cross-leveling and in the reasons underlying it. We then conducted a quantitative analysis of more comprehensive data on cross-leveling actions to measure the extent of cross-leveling for the Army RC, verify its relation to readiness shortfalls, and compare the results with the SMEs' reports of their particular experiences. This initial phase of the research is reviewed briefly in Chapter Two.

The next step in the research was to extend the analysis of readiness shortfalls and cross-leveling requirements into the future (Figure 1, upper right). Will there be problems in deploying urgently needed RC units after the drawdown? To find out, we built a model simulating personnel actions during mobilization. Specifically, this model allowed us to determine whether there might be shortfalls of qualified personnel serious enough to prevent the deployment of critical units at their prescribed strengths and timetables. It also determined the extent to which such shortfalls could be remedied by special train-up and cross-leveling actions. In Chapter Three, we describe the model and its application to a scenario entailing a major regional contingency (MRC).

To shed light on the ability of various policy alternatives to enhance peacetime RC personnel readiness and lessen the cross-leveling requirement, we built and ran a second model. Two principal policy approaches were examined, the first entailing an improvement in the utilization of active-duty experience within the RC and the second a reduction in the RC personnel turnover rate. The modeling results provided general guidance about the policies most effective at increasing job qualification and experience levels. The policies required to achieve such gains come at a cost: the resources needed to carry them out. The cost of these resources might, however, be partly or wholly offset by savings from reduced accession and training requirements. The model calculated those requirements as well, as we discuss in Chapter Four.

The costs and savings were calculated in two steps. First, we reviewed the relevant literature and programs to estimate the potential costs of policies to enhance readiness. In addition, we did a separate analysis of recruiting and training costs for the Army RC; the aim was to reduce the requirements generating these costs through the policy alternatives and use the resulting savings to help pay for the readiness enhancement policies. In the second step, we used these potential costs and savings as inputs to the readiness enhancement model. The model applied the costs incurred to achieve readiness enhancements under alternative policies and compared them with the savings resulting from reductions in annual accession and training requirements. These costs and savings are discussed in Chapter Five. Both the potential benefits and costs of the policy alternatives were taken into account explicitly in developing our final inferences and recommendations for policy. We determined which policies are likely to be most cost-effective and, from an affordability perspective, which policies may have wide applicability versus others that may have to be targeted more narrowly to units with especially critical shortages. This is discussed in more detail in Chapter Six.

Chapter Two

CROSS-LEVELING OF UNITS FOR
OPERATION DESERT SHIELD/STORM

We began our research by establishing the extent to which cross-leveling was required to address readiness problems during Operation Desert Shield/Storm (ODS/S). We conducted interviews with experts who were involved in the mobilization and deployment of RC personnel during ODS/S, particularly the personnel of the First Army region. These interviews had two purposes: (1) to learn what the experts experienced with respect to the extent of cross-leveling and the reasons for it and (2) to ascertain the availability of records of cross-leveling actions. We then acquired such records for the Army RC units activated during ODS/S and validated them by comparison with the reports and more limited, detailed records provided by the experts. Finally, using the records, we undertook a quantitative analysis of ODS/S cross-leveling actions for the Army RC.

INITIAL ASSESSMENT

We interviewed a number of different offices at U.S. Forces Command—then a joint command—and had detailed discussions with personnel who were especially knowledgeable about the readiness of the Reserve Components and the cross-leveling and mobilization actions that occurred during ODS/S. In particular, we met with personnel involved in the mobilization at the Dix and Aberdeen mobilization stations and held an extended series of discussions with representatives from the First Army and the 77th Army Reserve Command (ARCOM).

The SMEs indicated that there had been substantial cross-leveling during ODS/S. They reported that the goal was to fill the typical unit with a sufficient number of Duty MOS–qualified soldiers[1] to allow it to deploy to the mobilization station with a job qualification rate of 85 percent or better of required strength (C-1).[2] This figure was chosen to improve readiness to a wartime footing and, at a minimum, to try to ensure that the unit would maintain at least a C-3 rating after the potential loss of deployability status for some members upon further verification actions at the mobilization station. Indeed, it did prove necessary to undertake a small amount of additional cross-leveling at the mobilization stations during ODS/S; about 10 percent of the cross-leveling actions were carried out at the stations. Finally, there was consensus that the major factor underlying cross-leveling was a shortage of Duty MOS–qualified (DMOSQ) personnel.

DATA FOR COMPREHENSIVE ANALYSIS

To conduct a comprehensive quantitative analysis of cross-leveling throughout the Army RC during ODS/S, we obtained from the Defense Manpower Data Center two databases containing unit assignment information. The Selected Reserve database contained information on personnel serving in the RC just before the incursion into Kuwait. The activated-reservist database provided records for all reservists who were activated during ODS/S.

First, however, we validated this information by comparing it with two other sets of information: (1) very detailed paper-and-pencil records of cross-leveling actions and computer records of unit assignments maintained by the 77th ARCOM and (2) computerized databases developed by the First Army for its entire region, showing the compatibility between soldiers' initial units and the units in which they served in ODS/S. We believed the 77th ARCOM data in particular to form a nearly complete record of cross-leveling actions for those force elements.[3] After this validation we acquired infor-

---

[1]Personnel qualified for the military occupational specialty associated with their assigned job within their unit.

[2]As noted, C-3 requires a 65 percent DMOSQ level and C-1 an 85 percent DMOSQ level.

[3]To validate the RC-wide information, we compared the RC-wide database records for 77th ARCOM personnel with the 77th ARCOM and First Army records. Our analysis revealed that the RC-wide databases accurately represented cross-leveling actions.

mation from Forces Command on the required strength of RC units and their readiness levels, adding it to the RC-wide databases.

CHARACTERIZING ACTIVATED UNITS AND PERSONNEL

RC personnel and units activated for ODS/S may be characterized as shown in Table 1, which shows their status at the time of the Iraqi incursion into Kuwait. The table distinguishes personnel activated for ODS/S by the type of unit in which they served. It also shows that the overall fill rate of the activated units—the number of assigned personnel relative to required personnel—averaged about 89 percent. The DMOSQ rate was 70 percent in terms of assigned personnel, and about 63 percent against required strength.[4] As shown in Figure 2,

Table 1

Characteristics of Army RC Units Activated for ODS/S

Unit Characteristic	Percentage
Type of unit	
Combat service support	51.8
Combat support	25.7
Combat	12.5
TDA	10.0
Average fill rate	89.1
Average DMOSQ rate as percentage of	
Assigned personnel	70.3
Required strength	62.6
Average percentage of deploying personnel	
cross-leveled into unit	18.0

NOTE: Percentages are weighted to be representative of all personnel. Data are for July 1990 except for cross-leveling, which is aggregated over the final manning of the activated units.

The few discrepancies were resolved through analysis and discussions with the subject matter experts, in order to provide the best interpretation possible of the RC-wide data.

[4]This figure does not include a decrement for nondeployability, which probably amounts to about 6 percent of the requirement.

the shortage of DMOSQ personnel was accounted for more or less equally by the following:

- Unfilled positions.

- Positions filled with unit members who recently joined the RC and were in the process of receiving initial skill training in their DMOS.[5]

- Positions filled with persons qualified in a MOS other than their current DMOS.[6]

Finally, our analysis revealed that 18 percent of the reservists serving on active duty in operating RC units during ODS/S were cross-leveled into those operating units. In other words, about two of every

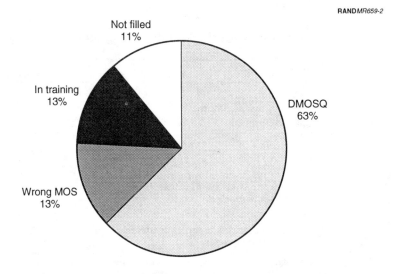

RAND*MR659-2*

Not filled
11%

In training
13%

DMOSQ
63%

Wrong MOS
13%

**Figure 2—Fill and DMOSQ Profile of Average Activated Army RC Unit
(Data for July 1990)**

---

[5]There is no Trainees, Transients, Holdees, and Students (TTHS) account for the RC as there is in the AC: Reservists are assigned to operational units and counted against unit strength while they are being trained.

[6]Retraining for current DMOSs is limited by school capacity, funds available, and the willingness of civilian employers to grant time off.

ten reservists serving in an operating unit during ODS/S were reassigned to that particular unit from another unit in the RC.

These data are generally consistent with the reports we received from our subject matter experts. The average fill rate corresponds closely to that reported by the SMEs, as does the DMOSQ level. As noted, the experts reported that the object of the cross-leveling was to deliver activated support units to the mobilization stations filled to 85 percent of the wartime requirement with DMOSQ personnel. If we take the DMOSQ rate shown in Table 1 (63 percent) and account for the unit members who completed training after July 1990 and subsequently deployed with their unit during ODS/S, we increase the DMOSQ level to about 70 percent of the requirement.[7] Accounting for the soldiers cross-leveled into the unit brings the DMOSQ level to 85 percent. (This is because a unit deploying at 85 percent DMOSQ with 18 percent (of the 85) representing cross-leveled soldiers would have a pre-cross-leveling DMOSQ rate of 70 percent.)

These results varied somewhat by component (USAR versus ARNG) and type of unit, but in each case the average unit was well short of the goal for deployment at the time of the incursion. Combat units showed the highest DMOSQ levels (about 75 percent), followed by transportation units (72 percent), other support units (62 percent), and, last, medical units (53 percent). Medical units had lower DMOSQ rates in part because they were intended to undergo extensive cross-leveling as part of the mobilization plan. For other units, the differences in DMOSQ levels generally reflected the corresponding differences in fill rates and C-rating requirements for these unit types. Among assigned personnel, the DMOSQ rates were similar (about 70–75 percent). The average fill rate and DMOSQ rate respectively were 6 and 8 percentage points higher for ARNG units than for USAR units. These differences generally reflected the presence of the combat units within the Guard.

This analysis thus suggests that cross-leveling to raise DMOSQ levels was important and extensive during ODS/S. A further verification of the relation between DMOSQ rates and cross-leveling is provided in Appendix A. It describes the results of a regression analysis we con-

---

[7]Such persons typically made up from one-half to two-thirds of the unit members in MOS training in July. See Appendix A for a more complete discussion.

ducted of cross-leveling, which shows the significant role of DMOSQ shortfalls in generating cross-leveling actions.

Low DMOSQ rates still are very much an issue. Defense Manpower Data Center records for the ARNG and USAR as of the beginning of FY94 as well as USARC monthly reports of DMOSQ rates for all USAR units show only a 4 to 5 percentage point gain since 1990. Beginning in FY94, unit closures and changes in the MOS distribution in the Army RC related to downsizing have exacerbated DMOSQ problems.

Chapter Three

PERSONNEL READINESS SHORTFALLS IN HIGH-PRIORITY UNITS

In this chapter, we use our mobilization model to explore the extent to which the readiness issues for the Army RC apply in particular to support units that will be required to deploy in future contingencies ("high-priority units"). We seek the answers to two related questions: (1) Are there high-priority units with critical shortages of DMOSQ personnel such that they will be prevented from deploying at their targeted strengths and timetables? (2) If so, how short are they?

To answer these questions, we began by using official records from the Army's input to the Mobility Requirements Study (MRS) to specify high-priority units and their deployment timetables for the Northeast Asia, Southwest Asia, and two nearly simultaneous MRC scenarios. We then modeled the process of attempting to fill these units to their targeted levels with DMOSQ personnel. Finally, we determined whether that attempt was successful, that is, if there were remaining DMOSQ problems that would prevent deployment of the units at their targeted levels.

The future structure of the RC cannot be known with certainty at this time. For our study, we optimistically assumed that the Reserve Components of the future were ideally structured for the Army's official two-MRC scenario. That is, we structured the RC to provide as many units of each type as were called for in the official (MRS) two-MRC deployment scenario. We filled out the remaining endstrength of the RC with the same types and proportions of units as those called for in the scenarios. Thus, both the units to be deployed and the additional units from which soldiers could be drawn for cross-leveling were perfectly matched to the two-MRC wartime scenario.

Since the official requirements for these scenarios are stated by Standard Requirement Code (SRC) rather than by unit, we matched RC units to the SRC list to generate a unit-level deployment requirement and timetable. Because there is no applicable official requirement, we excluded combat and special-operations forces and those organized according to a table of distribution and allowances (TDA). However, support forces in combat organizations were included for higher-echelon support when those organizations' SRCs were called for in the contingency plan. The resulting force represented just over half of the entire RC (233,000 of 460,000 enlisted personnel).[1] In selecting units for the contingency according to the list of SRC requirements, we first drew from Contingency Force Pool (CFP) units—those with the highest priority and readiness goals—before drawing randomly from other RC units. We subsequently used the personnel requirements, fill, DMOSQ and peacetime in-training rates of these actual units in our analysis.

The deployment timetable for these units is important, because it bears on whether one must cross-level only persons already qualified (prior to the mobilization) for the targeted DMOS in the receiving unit or, instead, whether there is sufficient time to train up soldiers in special (accelerated) wartime programs before cross-leveling them into the high-priority units. The timetable for deployment also has important implications for the desirable peacetime readiness level for a unit and, relatedly, the amount of cross-leveling to plan on for that unit during wartime; we will return to this point shortly.

We now discuss how we model the processes undertaken to bring the mobilized units as close as possible to 85 percent DMOSQ. Recall that we have matched actual units to the (required) SRC list for the scenario; the Duty MOS, Primary MOS, and training status of the soldiers assigned to these units provide the parameters used in the mobilization modeling. Within the mobilized units, the first step is to realign the personnel who already are there in order to produce

---

[1]This represents today's proportion of support units, which we have optimally configured to meet the two-MRC SRC requirements.

the maximum match between Primary MOS and Duty MOS. In other words, if a soldier is not qualified for his current Duty MOS but is qualified for an alternative Duty MOS in the unit that is vacant or filled by a non-DMOSQ soldier, then we reslot him into the job for which he is qualified.[2] Next, we take account of soldiers who are in the training pipeline for their Duty MOS; if they complete training before the deployment time for their unit, we allow them to deploy with the unit.[3] Last, we also allow soldiers to be sent to special courses for training, subject to two restrictions: (1) time must permit training for that particular MOS;[4] (2) the soldiers have to meet certain pay grade and MOS criteria in order to be qualified for the courses. The standards we use to determine course length and grade-MOS prerequisites correspond to the procedures used during ODS/S.

A roughly parallel set of actions can take place in nondeploying units, to provide soldiers who can be cross-leveled into mobilized units where they are needed.[5] First, we can cross-level soldiers into the mobilized units who already are qualified in the duty positions that we need to fill. Second, soldiers in other units who are in the training pipeline and who complete training for their DMOS can be cross-leveled afterward into units with shortages in the MOS for which they have just completed training. Last, and subject to the same restrictions described above, we can send soldiers to special training courses and then cross-level them into deploying units needing sol-

---

[2]The pay grade requirement for the position must be no more than one grade level up or two down from the reservist's current grade.

[3]Based on the training completion rates observed during ODS/S, we assumed that the reservists in the training pipeline would be trained up within one year or less; we thus assume that an additional one-twelfth of such soldiers became DMOSQ every month. Thus, for a unit deploying in 90 days, for example, we raise the number of DMOSQ soldiers in the unit at deployment by 25 percent (three-twelfths) of those in training on C-day.

[4]The period available for such training is limited by the time required for collective training in the high-priority unit prior to its deployment. In any event, we do not allow soldiers deploying within the first 45 days to be sent to such courses, because of the setup and training time involved.

[5]Cross-leveling of personnel *from* other high-priority units was permitted only if the unit had more than 100 percent of the required strength for a shortage MOS or had no requirement for a soldier's MOS.

diers in the MOSs for which they have been specially trained.[6] These processes are illustrated in Figure 3.

The results for the three scenarios are roughly similar in terms of the *percentage* of required units with readiness shortfalls (see Appendix B); of course, there is a greater *number* of units required for the two-MRC scenario than for the single MRCs and, thus, a greater total number with personnel readiness shortfalls. Below, we focus on the Southwest Asia MRC.

RAND*MR659-3*

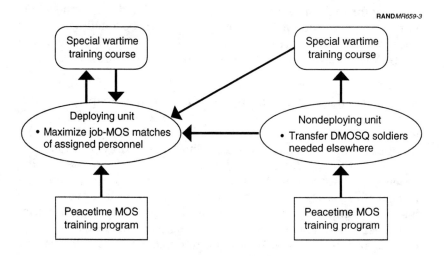

Figure 3—Processes Modeled to Bring Deploying Units to 85 Percent DMOSQ Level

[6]It is possible, but unlikely, that the simulated processes may be conservative with respect to readiness (i.e., they may overstate the number of personnel needed to fix the problems identified in high-priority units). This is because the DMOSQ figure is based on DMOS-PMOS matches. Our work suggests that about 15 percent of the mismatched personnel not in training for their DMOS may actually be qualified by virtue of secondary MOSs. In general, however, the effect of this secondary matching would be small, amounting to only 2–3 percent of the requirement. In contrast, the model optimizes both the cross-leveling and train-up processes. Soldiers are transferred to the unit that needs them the most. Train-up is perfectly scheduled to optimize the arrival of the newly trained soldier in the high-priority unit before it must deploy; moreover, constraints on retraining related to cost and infrastructure are not considered, and, as noted, immediate access to the soldier for retraining is presumed. These factors act to understate the personnel required to overcome shortfalls.

SIMULATION OF SOUTHWEST ASIA SCENARIO

Table 2 presents simulation results for the Southwest Asia scenario. Since, as noted, we perfectly configured the RC for two MRCs and are mobilizing for only one, this case is optimistic with respect to both force structure and deployment requirements. The results in the first row of the table indicate that if we also optimistically plan to rely on cross-leveling into all deploying units with DMOSQ shortfalls, the percentage of mobilized units that do not meet their DMOSQ targets for deployment after cross-leveling is small, about 1 percent. The percentage of MOSs involved is approximately 9 percent. That is, of all the specialties that need to deploy, about 9 percent fall short of the targeted DMOSQ level.

While the foregoing results provide some reassurance, we might think it sensible to make less optimistic assumptions. Consequently, we modeled the case in which we plan only for units deploying after C+30 days to cross-level. We choose this case because Mobilization Day may not be C-day; for example, it may be C+20, as it was in ODS/S, or even later. If so, accounting for the five days of preparation and collective training time assumed in the model based on ODS/S experience, we would be asking units scheduled to deploy within 30 days to be ready to deploy almost immediately upon mobilization, with little to no time to cross-level and integrate new personnel into the unit. More generally, while cross-leveling can be a cost-effective means to help ensure unit deployability, it is not the ideal solution to reserve readiness problems. The greater the reliance on cross-leveling to offset unit readiness shortfalls, the less the

Table 2

Prevalence of MOS Shortfalls Sufficient to Impede
Unit Deployment in Southwest Asia Scenario

Degree of Cross-Leveling Planned	Percent of MOSs Affected	Percent of Units Affected
Planned for all units	9.3	1.4
Planned only for units deploying after 30 days	30.2	25.4

NOTE: This scenario requires 566 units (containing 248 DMOSs).

likelihood that units will have had peacetime individual and collective training adequate to permit cohesive performance of their wartime mission. This is particularly true for units that must deploy early in the contingency. Finally, the reality is that the future force structure is not likely to be ideal, providing fewer of the needed units to deploy and fewer like units to feed them with cross-leveled personnel. Therefore, we need a more conservative case.

When cross-leveling is planned only for units deploying after 30 days, we get a very different picture. Now, one-quarter of all the units called for in the contingency are not able to deploy at the targeted DMOSQ level. As seen in Table 2, this involves about one-third of all the occupational specialties required for the contingency.

In Table 3, we show the distribution of the 25 percent of the units falling short of their targeted personnel readiness levels according to the severity of the shortage. The categories were determined according to how much improvement in the DMOSQ rate is needed to bring a unit up to the 85 percent goal for deployment to the mobilization station. For example, in the first row we see that an increase of less than 15 percent is needed; this means that the current DMOSQ rate in that unit—shown in italics—stands at about 75 percent of the requirement or better, because a relative improvement of less than 15 percent increases the DMOSQ rate to 85 percent or better. The "15 to 30 percent" row represents units having DMOSQ rates between 65 and 74 percent. The bottom row shows the percentage of units with DMOSQ levels below 65 percent of the requirement.

Table 3

DMOSQ Improvements Required to Eliminate Shortfalls Impeding Unit Deployment in Southwest Asia Scenario

Required Relative Increase in DMOSQ Rate *(current rate)*	Percentage of Units Affected
Less than 15 percent *(75%–84%)*	3.4
15 to 30 percent *(65%–74%)*	6.0
More than 30 percent *(below 65%)*	16.0
Total	25.4

NOTE: Assumes no cross-leveling for units deploying within 30 days.

We find that about 15 percent of the problem units (3 percent of all units) need increases of less than 15 percent. A somewhat greater number fall between 15 and 30 percent, and close to two-thirds fall above 30 percent. As will be discussed subsequently, these categories are important conceptually because (1) they correspond to C-ratings for personnel readiness and (2) the policies that we believe can be implemented to help eliminate DMOSQ shortages are likely to fix units needing an increase of up to 15 percent; cut the shortages by half or more for cases in the intermediate group, reducing them to fairly small numbers; and reduce shortages in the remaining units by up to 50 percent.

Chapter Four

POTENTIAL FOR ENHANCING ARMY RC READINESS

We demonstrated in Chapter Two that reserve readiness shortfalls, as measured by DMOSQ rates, have been large enough in the past to warrant significant cross-leveling. In Chapter Three, we showed that even if the future Army RC force structure were perfectly constructed to match the two-MRC scenario and were asked to respond only to a contingency in Southwest Asia, it would have problems if it lacked the ability to cross-level soldiers into the units deploying within the first 30 days of that contingency: shortages of qualified soldiers would prevent 25 percent of the Southwest Asia–contingency units from deploying at their targeted DMOSQ levels. We indicated that it would be wise to enhance the peacetime readiness of these units by raising their DMOSQ levels rather than relying on such eleventh-hour cross-leveling.

In this chapter, we ask what steps might be taken to raise peacetime DMOSQ levels. We examine the potential effects of several policy alternatives that might enhance peacetime personnel readiness. The results provide guidance concerning (1) the types of policies most effective at increasing job qualification and experience levels, and (2) their concomitant effects on cost-related factors, for example, annual accession and training requirements.[1] We begin by looking at underlying causes of DMOSQ problems; this analysis of causes provides the basis for identifying potentially effective policy approaches.

---

[1]Chapters Five and Six take cost into account explicitly: the cost of alternative incentives to enhance readiness weighed against their related savings in recruiting and training costs and their benefit in enhanced readiness.

UNDERLYING CAUSES OF READINESS SHORTFALLS

Why do the Army Reserve Components face significant readiness issues, in particular, a low rate of DMOSQ and a consequent requirement for substantial cross-leveling? A variety of reasons for readiness shortfalls have been identified by earlier research, much of it conducted within the National Defense Research Institute at RAND (NDRI, 1992; Buddin and Grissmer, 1994). These can be grouped into two principal causes—incomplete utilization of the active duty experience of Army personnel and high RC personnel turnover rates.

Incomplete Utilization of Active-Duty Experience

One aspect of the incomplete utilization of active experience is the low percentage of Army RC personnel with prior active duty service, relative to that in the other services (see NDRI, 1992). This was recognized in the Title XI legislation,[2] which required as a goal that 50 percent of enlisted ARNG members have at least two years of prior active duty service. With the exception of the Marine Corps Reserve, all the non-Army components are above the 50 percent prior service level. However, both Army components are well below 50 percent, with the Guard percentage falling in the mid-30s.

As noted in the NDRI analyses, the low Army Reserve and Guard figures are not the result of the Army's doing a poor job of bringing soldiers into the RC from its Active Component. Historically, the Army has brought into the RC better than one out of every three prior service soldiers, a rate exceeding those of the other services. The Army RC's prior service inventory percentage is lower because of the relatively small size of the Army Active Component relative to the Reserve Components. Historically, that ratio has stood at about one to one, whereas the ratio for the other services has generally been at least three to one. The substantially lower Army ratio provides much less leverage in drawing personnel with prior active duty experience into the RC. This limits the gains realized from the Army's record of drawing a higher percentage of prior service individuals into the RC. The problem is likely to worsen in the future, because the Army AC endstrength is to be drawn down more than the RC endstrength (to

---

[2]Army National Guard Combat Readiness Reform Act of 1992, P.L. 102-484.

495,000 versus 575,000, respectively; approximately 410,000 versus 460,000 enlisted personnel).

A second aspect of incomplete prior service experience utilization is that the former AC personnel who do join the RC often do so in a different MOS. For many soldiers this may reflect the selection of assignments near their priority geographical location, with less priority being given to the MOS vacancy requirement at recruitment. Recent data indicate that only 65 percent of persons entering the RC do so in their active duty specialty. That rate represents a dramatic improvement over the late 1980s level (which stood at about 40 percent). Nonetheless, some 35 percent of prior AC personnel entering the RC still do so in an occupation other than their active duty specialty. Thus, the benefit of their active duty specialty training and their experience in their MOS is lost, and training costs accrue.

High Personnel Turnover Rates

The second factor contributing to readiness shortfalls is personnel turnover: this includes substantial rates of job changes and attrition. The middle column of Table 4 shows the personnel turnover profile of the Army RC over the course of FY93, based on Selected Reserve records.[3] After one year, only 63 percent of the RC personnel are still in the same job. The remainder is split roughly equally between persons who have separated from the RC and those who are still in the RC but have changed jobs.

Table 4

FY93 Turnover in Army RC and Implications for DMOSQ Rate

Status After 12 Months	Percentage of Personnel	Percentage DMOSQ
Same job	63.2	86.6
New job	16.6	36.4
Not in RC	20.2	—

NOTE: Data report personnel turnover over the course of FY93.
SOURCE: Selected Reserve database.

---

[3]FY93 represents the most recent year approximating a steady state for the Army RC; since then, drawdown-related force restructuring actions have increased turnover.

The rightmost column of Table 4 shows the very significant effect that job changes have on the DMOSQ rate. The first row indicates that on average, nearly 87 percent of the persons who have not changed jobs within the last 12 months are DMOSQ by the end of that period; only 13 percent still need training.[4] There is a marked contrast, however, for persons who change jobs. Their qualification rate falls to only 36 percent. Of course, the benefit of training and experience is lost for persons who separate.[5] Most such soldiers— who account for 20 percent of the cohort by the 12-month point— must be replaced by non-DMOSQ persons requiring initial or entry retraining. These results make it very clear that personnel turnover has a substantial negative impact on Duty MOS qualification rates, increasing the requirement for cross-leveling.[6]

The Arroyo Center's recent work on the Army's FY92–93 BOLD SHIFT training initiative (Sortor et al., 1994) and on redesigning the Army school system (Winkler, 1995) confirms the existence of a high rate of personnel turnover in the Army RC and its serious consequences for unit readiness. The conclusions of these studies emphasized the importance of addressing turnover if real progress is to be made in improving training and readiness. The researchers identified a number of related factors underlying readiness shortfalls, including DMOSQ shortfalls, difficulty in scheduling many of the individuals in the RC needing MOS training for their individual training programs, and the conflict between individual and collective training needs.

EFFECTS OF SPECIFIC PERSONNEL MANAGEMENT CHANGES

We now seek the best way to enhance future peacetime personnel readiness levels. To help achieve this objective, we built a model that allows us to simulate peacetime personnel characteristics and readiness levels of the postdrawdown Army RC force under current and

---

[4]MOS training in the RC often is split over two years; thus, some soldiers need to remain in the same DMOS for more than one year to become DMOSQ.

[5]Some of these individuals do eventually return in the same MOS.

[6]These figures are for the Army RC as a whole; in recent times—e.g., FY93—enlisted personnel turnover in the USAR has been about 8 to 9 percentage points greater than in the ARNG.

alternative personnel policies. Given the foregoing results, we were particularly interested in the readiness enhancement effects of:

- Increased levels of prior active service experience (through increased accessions and MOS matching)

- Reductions in personnel turnover (job changes and attrition).[7]

We wanted to look at the impact of improvements in the above factors on two key readiness dimensions:

- Job qualification levels (DMOSQ rate and depth of experience)

- Annual accession and training requirements.

As noted earlier, the cost of the policies needed to secure the enhancements might be offset in whole or in part by reductions in recruiting and training requirements. These costs and savings are estimated in Chapter Five.

The Readiness Enhancement Model

The readiness enhancement model projects the future (postdrawdown) Army RC enlisted inventory in peacetime. In the model, the RC personnel inventory is categorized along four dimensions:

- Personnel with two or more years of prior Active Component service versus those without.

- Pay grade. E1 through E3 are tracked as one category, as are E7 through E9. E4, E5, and E6 are each tracked separately.

- Personnel accessing within the last 12 months ("new" personnel), as opposed to those with at least a year behind them ("old"). (Turnover behavior and job qualification rates of first-year and

---

[7]We take the programmed endstrengths of the AC and RC as given. Of course, a different approach to solving the problem would be to shift force structure from the reserves to the Active Component or to increase the size of the RC—increasing accessions into critical units and MOSs—while holding constant the reserve mission in MRCs. Such policies that change endstrengths, which would require congressional action, are outside the scope of this study.

more experienced personnel differ enough to warrant making this distinction.)

- DMOSQ personnel versus not-DMOSQ personnel.

The total number of inventory categories is thus 2 (prior active service versus non-prior active service) × 5 (grade categories) × 2 (old versus new) × 2 (DMOSQ or not), or 40.

Personnel are subject to four events annually:

- They can attrite from the force, in which case they are subtracted from whatever inventory category they are in and not added elsewhere. (The model assumes that about one-third of separated personnel eventually return to the Army RC.) Those who do not attrite can undergo one or more of the other three events.

- They can be promoted from one grade category to the next.

- They can change jobs. (This does not by itself involve a change in inventory category.)

- They can change from not-DMOSQ to DMOSQ, or (most likely because of a job change) vice versa.

Of course, many personnel do not undergo any status change during an annual cycle—they remain in the RC in the same job, same grade, and same qualification status. At the other extreme, an individual remaining in the RC could change status in all three other categories. Transitions are illustrated by the matrix in Figure 4, which covers, in the row stubs at the left, all the inventory categories that "new" or "old" reservists could occupy. In the cells, we put transition probabilities governing how personnel change states during an annual cycle, including the simultaneous change of two or three states and the possibility of remaining unchanged. Appendix C presents the full set of transition probabilities (Table C.2) and discusses their derivation from Selected Reserve data. There is a probability for attrition and a probability for each of the eight possible combinations of the other three events. For each of the inventory categories, these nine probabilities sum to one. Multiplying the transition probabilities by the start-of-year inventory for each row yields personnel flows between categories, including attrition and number of job changes.

Current Status	Attrited	Status One Year Later							
		Not promoted				Promoted			
		In same job		Different job		In same job		Different job	
		NQ	DMOSQ	NQ	DMOSQ	NQ	DMOSQ	NQ	DMOSQ
Non-prior active service									
E1–E3 not DMOSQ									
E1–E3 DMOSQ									
E4 not DMOSQ									
E4 DMOSQ									
E5 not DMOSQ									
E5 DMOSQ									
E6 not DMOSQ									
E6 DMOSQ									
E7–E9 not DMOSQ									
E7–E9 DMOSQ									
Prior active service									
E1–E3 not DMOSQ									
E1–E3 DMOSQ									
E4 not DMOSQ									
E4 DMOSQ									
E5 not DMOSQ									
E5 DMOSQ									
E6 not DMOSQ									
E6 DMOSQ									
E7–E9 not DMOSQ									
E7–E9 DMOSQ									

NOTE: "NQ" means not DMOSQ.

Figure 4—Transition Matrix

To simulate the effects of an increase in prior active service inventory, for example, we augmented the number of accessions in the prior service category (the bottom half of the figure), decrementing correspondingly the non-prior active service accessions, and determined the steady-state force composition. To look at the influence of improving MOS matches among the prior active service soldiers entering the RC, we increased the number of DMOSQ accessions among prior service personnel and decremented not-DMOSQ accessions accordingly. By analogy, to look at the effect of cutting attrition, we decreased the cases in the "Attrited" column and redistributed them proportionately over the remaining columns. To look at the effect of cutting the rate of job changes, we reduced the cases for the "Different job" columns and redistributed them to the corresponding "In same job" columns, in proportion to the NQ versus DMOSQ distributions for those columns. (For example, "Not promoted, Different job" reductions were redistributed to the "Not promoted, In same job" cells, in proportion to the NQ-DMOSQ ratio for the "Not promoted, In same job" cells.)

The "base case" postdrawdown RC personnel inventory is derived from a scaled-down version of the FY89 Reserve Component inventory (776,000 to 575,000), as reported in the Selected Reserve files. Although more recent inventory information was available, we believe it was affected by the AC drawdown in ways that make it inappropriate for use in steady-state postdrawdown modeling. For example, beginning in FY90, Army AC accessions were cut well below the level required to sustain the force in order to reduce the number of older service members who had to be induced to leave. This increased the seniority of the active force grade structure. Accessions into the AC now are beginning to rise to their required levels, but the new steady state will not be reached for another four to five years. Thus, post-FY89 figures on RC inventory would be affected by drawdown-related changes in accessions into the RC from the Active Component, and would not represent the true future flows from the AC into the (postdrawdown) RC force in terms of overall numbers or distribution among pay grades.[8]

---

[8]We adjust for two important differences between the FY89 and postdrawdown AC flows into the RC. We apply a factor less than one to the "new" (accession) prior active service inventory categories to reflect the reduction in supply once the active force is

In generating transition probabilities, we tried to approximate recent experience while maintaining consistency with the anticipated steady-state nature of the postdrawdown force. The initial transition probabilities we used were the ones prevailing in the Army RC at the beginning of FY94 (i.e., those describing soldiers' transitions during the 12-month period of FY93). As noted, we used these probabilities because RC downsizing actions initiated in FY94 to reduce, restructure, and realign units between the USAR and ARNG have for the moment increased the rate of MOS changes and separations, making more recent transition information unrepresentative of the future steady state.

We applied the FY93 transition probabilities to the postdrawdown force structure in the model. That resulted in grade structure and endstrength changes, so we fine-tuned the promotion and attrition probabilities until they did not cause changes. (Appendix C gives the adjustment factors we applied to the FY93 probabilities to generate a steady-state postdrawdown force, the "base case.") We used the adjusted transition probabilities to determine the effects of policy changes.

The model assumes total endstrengths of 495,000 and 575,000 for the AC and RC, respectively, amounting to approximately 410,000 and 460,000 enlisted personnel. In assessing the effects of different policy approaches, the model preserves the RC endstrength and, to the extent possible, the base case RC grade structure and proportion of soldiers with prior AC experience; this is accomplished by adjusting accessions and promotions as needed.

To fully account for the potential benefits of prospective policies, the model maximizes the number of DMOSQ and prior active service personnel among those accessing into the RC. This is accomplished by imposing an accession hierarchy on the base case and all alternatives; it draws from the pool of potential accessions (supply) according to the following set of priorities: first, prior active service accessions to a MOS in which they are qualified; then, non-prior active service accessions to a MOS in which they are qualified (returning re-

---

downsized. By analogy, the proportion of such accessions that is DMOSQ is adjusted upward—and not-DMOSQ, downward—to reflect the recent improvements in MOS matching noted earlier.

servists); prior active service accessions to a new MOS;[9] and, finally, non-prior active service accessions to a new MOS.[10]

The model's output is the number of personnel in each inventory category under alternative assumptions about (1) prior active service accession rates into the RC and (2) their MOS match rate, (3) job change rates ("job turbulence") in the RC, and (4) attrition. This includes changes in accessions that may be required by preservation of endstrength—both by grade and in total. Concomitant changes in promotion rates also form part of the output. When totaled across the entire force, the category-specific changes in personnel inventory along with the category's annual event rates for attrition and job changes under the prospective policy result in new levels of DMOSQ and job experience.

The job experience measure we use, "Full-time Equivalent Job Years" (FEJY), represents the years of experience in the soldier's current DMOS, and is a variant of the "Full-time Equivalent Training Years" (FETY) measure developed by Grissmer et al. (NDRI, 1992). It is computed by dividing the total number of days the individual has spent in the DMOS by the peacetime number of annual work days in the AC (225). The total number of days was determined by simulating personnel flows through the postdrawdown RC base case—accounting for 38 days of RC duty per year, inventory category-specific job change rates in the RC, and prior experience among persons entering from the AC. At entry to the RC, prior service soldiers are credited 225 days times the number of years served on active duty if their AC MOS matches their RC DMOS, and 0 otherwise.

Changes in accessions, turbulence, and attrition rates also imply changes in training load. We assume that all accessions into the RC without prior military experience receive the same Advanced Individual Training (AIT) at AC installations received by their active duty

---

[9]When the policy is to increase prior active service accessions, the second and third priorities are switched.

[10]In the simulation, the number and makeup of accessions (the "new" inventory) is permitted to change for the first nine iterations in order to meet the constraints on endstrength and prior service inventory. After nine iterations, most of the changes have been realized, and the model then estimates the steady-state RC inventory using these accession results out to 30 years (iterations). Shukiar (forthcoming) describes the Readiness Enhancement Model in detail.

counterparts, an assumption borne out by our examination of recent SIDPERS and ATRRS records. (See Winkler (1995) for a description of these records.) Persons without prior military experience who join the RC also receive basic training (BT). Most persons previously in the military who enter the RC in different jobs or soldiers currently in the RC who change jobs must be retrained in their new DMOS. We use data derived from SIDPERS and ATRRS to allocate these new DMOS training loads between AC AIT schools and Reserve Component Training Institutions (RCTIs).[11]

Effects of 50 Percent Improvements in Selected Factors

The results depicted in Table 5 represent the estimated improvement in DMOSQ rate, job experience, accession requirements, and training loads resulting from a 50 percent improvement in selected fac-

Table 5

**Estimated Effects of 50 Percent Improvement in
AC Experience Use and in Turnover**

Policy Variable Improved	DMOSQ Rate (%)	Job Experience (yrs)	Accessions (000s)	Training Load (000s)		
				BT	AIT	RCTI
None (base case)	68.4	1.00	93.7	50.9	59.4	67.5
PS inventory	68.5	1.11	94.8	46.6	55.8	70.2
PS MOS match	71.0	1.22	93.6	51.1	58.3	59.1
Job turbulence	74.5	1.38	92.6	51.2	55.0	36.4
Attrition	73.6	1.03	46.6	26.7	35.9	65.2
Total turnover	79.7	1.45	45.8	28.2	32.8	34.7

NOTE: DMOSQ = Duty MOS qualified. The job experience variable represents the total duty days spent by the average soldier in the RC inventory in his current DMOS; he receives one year of credit for every 225 duty days spent in that job, but restarts at 0 days with every job change to a different occupational specialty. BT = Basic Training; AIT = Advanced Individual Training; RCTI = Reserve Component Training Institution.

---

[11]Accounting for soldiers who already are qualified for their new duty positions (and those not trained for their old ones), the available data suggest that about 87 percent of the job changes result in a *retraining* requirement. We use this figure in estimating training loads.

tors of the two principal types identified earlier: the utilization of active duty experience and personnel turnover within the RC. The 50 percent figure is intended to represent an upper limit on the amount of improvement that may be feasible.[12] It is important to bear in mind that the results are only estimates, and that they rest on a number of assumptions embedded in the model. Thus, the *absolute level* of a particular value, such as annual accessions, should be considered approximate.

Base case. The first row of Table 5 shows the base case (without improvements). These data indicate that a steady-state enlisted RC force of 460,000 personnel would be characterized by a Duty MOS qualification rate of just over 68 percent. The average number of (full time equivalent) years of job experience would be about one year. That is, the average Army reservist would have accumulated over the course of his career the same number of duty days in his current DMOS as if he had been on active duty in that DMOS for one year. The last four columns show annual accession and training requirements, in units of thousands of soldiers. For example, the number of annual accessions for the RC would be just under 94,000. The associated annual training load deriving both from initial entry training and from retraining persons who change jobs would consist of approximately 51,000 in BT, 59,000 in AIT, and another 67,000 in RCTIs.

Increased levels of prior active service experience. The second row indicates how those numbers would change if it were possible to increase the prior active service inventory level for the entire Army RC to 50 percent, as prescribed for the Army National Guard under Title XI.[13] In this example, this is accomplished by increasing the rate of prior service accessions in grades E4 to E9 up to 30 percent from the current rate—though the typical increase is much smaller—and in-

---

[12]Another approach would be to assess the effects of equal-cost improvements. Here, we assess costs separately (Chapter Five). We prefer this approach because the potential of the alternative policies to enhance personnel readiness in the Army RC varies greatly, independent of cost. These differences are noteworthy in their own right, and set the stage for limiting the discussion of costs to the most promising alternatives.

[13]According to recent data, this would represent approximately a 50 percent increase in the Army National Guard enlisted prior service inventory from its historical levels. Given the endstrengths used in the model, the overall increase in prior active service inventory across both reserve components is about 32 percent.

creasing prior service accessions in grades E1 to E3 to about 3.5 times their current level. These different ratios are required to preserve the grade structure. Because prior service personnel in the lower grades make up a smaller proportion of accessions than they do at more senior grades, they require greater augmentation relative to their base case accession level; moreover, such persons leave the RC at greater rates than do non-prior-service individuals, while this is less true of soldiers in grades E4 to E9.

The results indicate that even such a substantial increase in the percentage of the RC inventory with prior active service would produce little change from the base case. The DMOSQ rate rises by about 0.1 percentage points (less than 1 percent in relative terms). The increase in average number of years on the job is larger, but still amounts only to 0.11 years (11 percent). Because we are trading prior active service accessions for non-prior active accessions, the *total number* of annual accessions remains relatively constant. Prior service personnel need not receive Basic Training or initial entry training in AIT courses, so these numbers decline, by about 9 percent and 6 percent, respectively. (The AIT savings is smaller because, unlike BT, the program is given to some prior service accessions into the reserve—a portion of those retraining in a new MOS.) In contrast, the increase in prior service accessions causes the RCTI load to rise by about 4 percent. This reflects retraining of soldiers who enter the RC in different jobs from those held in prior active service.

Why the limited benefits? First, to live within the grade and total endstrength constraints while increasing prior active service accessions requires accepting some prior active personnel in mismatched MOSs in lieu of accessing returning non-prior active service reservists qualified in those MOSs. This lowers DMOSQ rates and job experience. Second, the high rate of personnel turnover in the RC eventually leads to attrition or job changes (and resulting loss of job qualification and experience) among many of the prior active service personnel.

The third row addresses the effects of increasing the MOS match rate for persons who enter the RC from the Active Component. It represents a 50 percent relative improvement in the DMOS match rate of prior active service accessions, from 65 percent to nearly 100 percent. As is true for increasing the prior active service inventory in the RC, these results indicate that even a nearly perfect MOS match rate would provide only modest readiness enhancement from the base case. Again, personnel turnover eventually takes its toll.

The DMOSQ rate rises by less than 4 percent. Since we are not reducing non-prior active service inventory in the RC, the annual accession, BT, and AIT requirements remain essentially the same as for the base case. Because soldiers entering the RC from the AC now do so in the same MOS, the job experience level does show marked improvement (about 20 percent), and the RCTI load is reduced by lowering entry retraining requirements.

Decreased levels of personnel turnover. The remaining rows of Table 5 deal with the effects of reducing personnel turnover. They show considerably greater potential enhancement of RC personnel readiness. Row four estimates the effects of reducing job turbulence, that is, the rate of persons changing jobs within a given year, by 50 percent. It indicates a substantial effect on the Duty MOS qualification rate: approximately six percentage points (about 9 percent in relative terms). The job experience level shows marked improvement as compared with the base case (about 40 percent). Both these effects are on the order of twice those of the near-universal prior service MOS match. Of course, as attrition is not addressed, there is essentially no effect on accessions or, relatedly, on BT requirements. In contrast, because of the reduction in job changes, there is a large reduction in the RCTI load—down by nearly one-half. AIT declines, but by a much smaller amount. This is because the RCTI load is driven by retraining requirements, whereas the AIT load is driven primarily by initial training, and the BT load by initial training only.

In the fifth row, we see the estimated effects of lowering attrition by 50 percent. Again, there is a substantial increase in the DMOSQ rate, on the order of five percentage points. We do not see much change in the average years of job experience, which increases by less than 5 percent. The benefit in FEJY is small because, with the decrease in attrition, fewer accessions are required and there simply is not room to take all the available soldiers with prior military experience in a matching MOS. Because attrition is halved, the drop in accession requirements and, relatedly, in BT is about half. The annual AIT load—driven in part by retraining requirements—declines by 40 percent. Compared with the base case, the RCTI load goes down slightly (3 percent), because the smaller accession requirement generates a slightly smaller requirement for entry retraining.

The last row estimates the effects of reducing total personnel turnover by 50 percent. This means combining the 50 percent reduction in job turbulence with the 50 percent reduction in the attrition rate. It represents an interesting possibility, because one can conceive of similar policies to effect both types of changes. For example, that policy could involve paying bonuses to soldiers who remain in the RC in the same job. Moreover, the approaches should be complementary in their effects. Since the attrition and job turbulence changes address different types of turnover problems, the impact of reducing total turnover should be close to the sum of their independent effects.

The final row in Table 5 clearly illustrates this pattern. As compared with the base case, reducing total turnover by 50 percent provides a very substantial improvement in the DMOSQ rate to nearly 80 percent (a relative increase of nearly 17 percent); this is twice the improvement resulting from reducing job turbulence or attrition alone. It also captures the job experience increase (about 45 percent in total) and reduction in RCTI training requirements (about 50 percent) from turbulence reduction. Accession, Basic Training, and AIT savings related to reduced attrition also are preserved (about 45 to 50 percent).

In sum, the estimates presented in Table 5 illustrate the potential effectiveness of alternative approaches to enhancing RC personnel readiness. Although they rest on important assumptions, they clearly imply that among the options evaluated, job turbulence and attrition reduction, and in particular the combination of the two, promise by far the biggest improvements in DMOSQ rates and experience levels. They would also bring about the largest reduction in accession and training requirements, which might help to offset the costs associated with readiness enhancement incentives, such as cash bonuses. The potential improvement in these RC readiness factors appears to be much larger than that possible from increasing prior service inventory or MOS matching.

Chapter Five

COSTS AND SAVINGS OF READINESS ENHANCEMENT POLICIES

As we have seen, reductions in job turbulence and attrition appear promising for addressing shortages of DMOSQ personnel and of experience in the Army RC, along with the associated need for cross-leveling, which could threaten future deployments. What policies might accomplish such reductions? A variety of studies in both the Active and Reserve Components have demonstrated the powerful effects of economic incentives on a diverse set of desired behaviors, including enlistment, retention, reenlistment, and skill channeling into hard-to-fill occupational specialties. (See, for example, Buddin and Roan (1994), Fernandez (1982), Polich et al. (1986), and Hosek (1985).) As noted earlier, such incentives might reduce both job turbulence and attrition in the RC, and have been suggested by previous RAND analyses (for example, Buddin and Grissmer (1994), Sortor et al. (1994), and NDRI (1992)). The policies focus on the types of options listed below. We assume that it is not feasible to alter pay tables differentially for the RC relative to the AC; thus, augmenting compensation within pay grade would be accomplished by bonuses.

- **Raise compensation based on time in the job.** This must be part of the overall equation, since any attempt to control job turbulence on strictly administrative grounds is likely to increase rather than decrease attrition, particularly at the junior NCO and lower grade levels, with the resulting loss of much of the benefits seen in the present analysis. It is generally believed that many job changes are the result of promotion-seeking (Buddin and Grissmer, 1994). The amount of the annual bonus to discourage switching jobs could be a function of the increase in pay resulting from promotion to the next grade times the (prebonus) probability of obtaining that promotion within the next year given the soldier's MOS and time in grade. Because they increase compensation, such bonuses also should have some benefit in reducing attrition among those individuals considering leaving the RC.[1]

- **Expand reenlistment bonuses and link these bonuses to remaining in the same MOS.** These bonuses (or portions thereof) might be paid annually to eligible individuals who complete the year and who remain in the same job, providing both attrition- and turbulence-reduction benefits. If such policies are to be successful, however, it also will be necessary to compensate loss of promotion opportunity caused by increased retention.[2]

We recognize the need to offset the cost of increased compensation, particularly in the current budgetary environment. One strategy for accomplishing this is to attempt to offset those costs with the savings achieved through the reductions in accession and training requirements demonstrated above. We now review available information on those savings. Next, we estimate the cost of reducing reserve attrition and job turbulence through the measures suggested above. Finally, we weigh the costs of those measures against the savings.

---

[1] A variation on this policy to link pay to time in the job is to allow overgrading, i.e., promote individuals and allow them to stay in jobs conventionally assigned to lower ranks. In this case, reservists are promoted and receive the pay of the next-higher grade rather than being compensated for lowering their chances of promotion. An advantage of overgrading is that soldiers get the satisfaction and benefits of the promotion. A disadvantage is that the grade structure of the RC becomes more senior, which increases costs more than the simple increase in pay as well as making RC personnel more senior than active duty soldiers (with more experience) performing the same jobs. The latter issue could pose morale problems if AC and RC soldiers have to serve together and might even reduce AC retention by increasing flows to the RC.

[2] These bonuses apply to reenlistments among soldiers currently serving in a given RC unit and DMOS. Past research has shown that three-quarters of the soldiers who change Army RC units move to a unit located within five miles of their previous unit (Buddin and Grissmer, 1994). In other instances of unit changes involving large distances that are driven by the geographical relocation of the soldier's residence, it may be helpful to consider compensation for commuting overhead if the soldier agrees to remain in his previous unit and job.

RECRUITING SAVINGS

If Army reserve attrition declined, the USAR and ARNG would need to recruit fewer enlistees. Hence one implication of a policy that would reduce attrition is a cost savings in terms of lower recruiting costs needed to sustain the force. We will now review findings on the cost of recruiting to assess the approximate magnitude of this potential savings.

Research by Tan (1991) and Asch and Dertouzos (1994) provides estimates of reserve recruiting elasticities and costs. The studies note that recruiter costs are the primary resource costs of recruiting. Tan finds a non-prior-service recruiter elasticity—the percentage change in the number of recruits for a 1 percent change in the number of recruiters—that approximates that found for the regular Army (Goldberg, 1985); since several studies—including Polich et al. (1986)—found that the cost of recruiting someone into the active forces amounted to about $7,000 in FY91 dollars, Asch and Dertouzos estimated the cost of recruiting a non-prior-service *reservist* at $7,000 as well (about $7,750 in today's dollars).

Based on Tan (1991), we expect that the costs of recruiting prior service and non-prior-service reservists will differ. In models of recruit supply, Tan found that approximately the same level of recruiter effort was required to attract three to four prior service recruits as was needed to attract one non-prior-service recruit.[3] Using this information, we can modify the estimate that Asch and Dertouzos proposed for the cost of recruiting a non-prior-service reservist to produce an estimate for the cost of recruiting prior service personnel. This yields a recruiting cost of approximately $2,200 for a prior service reservist when adjusted for inflation.

TRAINING SAVINGS

We now summarize the methodology for calculating dollar savings from reductions in training load. (For a more detailed treatment, see Appendix D.) We first compiled the incremental cost of *MOS reclas-*

---

[3]This difference presumably arises from the relative ease of recruiting from a pool of people who have already shown some "taste" for the military, as opposed to recruiting from the general population.

sification training conducted in Advanced Individual Training (AIT) courses at active schools. These costs included those of the military personnel involved (school staff and instructors), procurement of any ammunition or other supplies and materials required, and base operating support. For most cost elements, we used TRADOC published factors to calculate the variable cost per course of changing training load. Student pay and allowances, one of two cost elements not covered by the TRADOC handbook, was computed by applying pay and allowance factors from the Force and Organization Cost Estimating System (FORCES) and the *Reserve Forces Almanac* to the length of individual courses. The cost of ammunition for each course was obtained from the Army Manpower Cost System (AMCOS) data file, maintained by the Army Cost and Economic Analysis Center (CEAC).

Of course, the cost of training for different MOSs varies considerably depending on the length and resource demands of the course. To consolidate to a single set of factors, we next calculated a weighted average course cost for each reserve component. The weights for individual course costs were determined by the size of FY93 E1 through E4 authorizations by MOS. Authorizations, in this instance, represent a surrogate measure of training requirements by MOS. We further assumed that the higher-demand MOSs—those accounting for 90 percent of authorized positions—adequately represented all MOSs.

After completing cost factors for active schools where AIT courses are taught, we compiled a parallel set of factors for reserve schools teaching reserve courses. In fact, the majority of reservists requiring reclassification training—those with prior (active or reserve) service in other MOSs—are retrained in Reserve Component Training Institutions (RCTIs), in courses that are reconfigured to comply with their 38-day-per-year training schedule. To estimate reserve school factors, we took into account that the RC-configured Program of Instruction (POI) trains soldiers in fewer training days than the AC POI; we estimated that the average RC course involves slightly less than one-third the days of the average AC course. Further, we incorporated into our cost factors the unique staffing structure of RCTIs. Because RCTIs are designed to provide training near the reservist's home station, they tend to be multifunctional rather than specialized, and smaller but more numerous than active schools. The three major types of RCTIs providing MOS training are U.S. Army Reserve Forces schools (USARFs), State Military Academies (SMAs) in the Guard, and more specialized Reserve Training Sites-Maintenance (RTSMs). Our factor for school staff pay and allowances in RCTIs represented a weighted average of the costs for the three major school types. The weights in this case were determined by the FY94 student load of the three types of schools.

Table 6 shows detailed AIT and RCTI course factors calculated by the methods described above. Overall, the RCTI MOS reclassification course incremental cost is slightly less than half the cost of the corresponding AIT course. Most of the factors directly reflect the RCTI course's shorter length. Student pay equals an amount halfway between the pay of an E4 and an E5, the average pay grade level of soldiers in reclassification training. Only the cost of school staff pay and allowances is higher for RCTI courses than AIT courses. This reflects the greater fixed cost of manning a dispersed system of schools, one that must give up the economies of scale characterizing the larger active schools in order to bring the training closer to the student.

For both types of courses in Table 6, we use the total cost number as the per-soldier savings from a reduction in training load. Table 7 shows those factors in the right-hand columns (rounded to the nearest $50), but also adds numbers for two other types of training: "Basic

Table 6

Average Course Cost Per Student for MOS Reclassification Training

Type of Cost	AIT Course ($)	RCTI Course ($)
Direct cost		
School staff pay and allowances	1,878	2,283
Student pay and allowances	7,242	2,087
Civilian pay at school	241	217
Other Operations and Maintenance (O&M)	102	46
Ammunition for training	253	115
Indirect cost		
Instl. staff pay and allowances	102	33
Installation support Operations and Maintenance (O&M)	380	123
Total cost	10,198	4,903

Table 7

Per-Soldier Savings by Type of Training

Basic Training ($)	AIT after BT ($)	AIT for Reclass. ($)	RCTI ($)
6,150	7,750	10,200	4,900

Training" and "AIT after BT." The difference in the two "AIT" columns (second and third columns) is purely a function of student pay and allowances: Soldiers receiving *initial MOS training* in the course (immediately following Basic Training) typically hold grade E2 and are paid accordingly; soldiers receiving MOS *reclassification training* receive pay between that of an E4 and E5 on average, as noted. Costs for Basic Training are estimated based on the methodology described in the *TRADOC Resource Factor Handbook* (p. 22 of the August 1994 version). We will use the information in Table 7 later in this chapter to estimate the cost savings resulting from alternative readiness enhancement policies.[4] For a more detailed breakdown of the savings by cost element, see Appendix D.

COST OF REDUCING PERSONNEL TURNOVER IN THE RESERVE COMPONENTS

Reducing Attrition

It is evident from the analyses presented in Chapter Four and earlier in this chapter that reducing reserve attrition—and, thereby, acces-

---

[4]The costs shown in Table 7 are the total incremental dollar costs of training an additional reservist and, thus, the savings when the training burden is reduced by one. Another approach, shown in Appendix D, calculates the *current net dollar savings* of reducing the training burden. The latter approach recognizes that in the current system some of the savings will be realized in nondollar terms. For example, today's reservists often must forgo unit training in order to attend school; in such cases, reducing the MOS training burden would increase unit readiness by freeing soldiers to attend unit training activities, rather than yielding net savings. However, in this chapter we express the full incremental savings in dollars. We assume that in the interest of enhancing readiness, current obstacles to providing IDT and AT to MOS trainees are overcome (or that the trainees are replaced during this time with DMOSQ personnel for the purpose of ensuring adequate collective training opportunities for the other unit members).

sion and training requirements—is likely to lead to substantial cost savings. However, we also expect that it would incur considerable outlays due to the required payment of incentives. We now summarize findings on the size of outlays required to induce declines in attrition.[5]

The Army's principal method for reducing reserve attrition has been to offer reenlistment bonuses. However, one could also view reserve wages as an incentive not to separate. Therefore, we consider research on both the relationship between reenlistment bonuses and attrition and the sensitivity of reserve manpower supply to reserve earnings.

The first military compensation test authorized by Congress studied the effects of reenlistment bonus payments on Army reserve reenlistment (Grissmer et al., 1982). While the results of the test are informative in assessing the outcome of the particular reenlistment bonus offered, the design was not sufficiently general to allow the results to answer many questions about the efficacy of other reenlistment bonus options. The test authorized bonuses to be offered only to reservists with no prior military service, those with greater than eight years of service at the end of their term, and those who were enlisted as of October 1, 1977. Hence, the results of the experiment are for a subset of reservists who may not be representative of reservists as a whole. In addition, only two types of bonuses could be offered: $900 for a three-year reenlistment or $1,800 for a six-year reenlistment, both to be paid half at the time of reenlistment and the other half in installments at the end of each year of service. It is unclear how well the results for these specific bonuses provide information about other bonus amounts with other enlistment term restrictions.

Using multivariate probability-of-reenlistment analysis based on this experiment, Grissmer et al. found that the bonus coefficient was statistically significant, but increased the reenlistment rate from 38.4

---

[5]Note that "attrition" is the fraction of individuals who leave the reserves at any point in their term of service, including those who leave at the end of their term. We use the term "retention" to refer to the fraction of individuals who remain in the reserves, so the retention rate is equal to one minus the attrition rate. A related concept is "reenlistment," which is the fraction of those individuals making it to the end of their term of service who commit to another term of service.

percent to only 40.6 percent. The effect on man-years was larger; the bonus substantially raised the average term chosen by those reenlisting, with bonus recipients choosing an average term of 4.37 years compared to 1.31 years for nonrecipients. Note that the experimental bonuses of $900 and $1,800 would compare to about $2,000 and $4,000, respectively, in today's dollars.

Using the same data analyzed by Grissmer et al., Burright et al. (1982) estimate reenlistment pay elasticity—the percentage change in the reenlistment rate due to a 1 percent change in reserve pay. They find a relatively small Army National Guard pay elasticity of 0.18.

In another study, Marquis and Kirby (1989a) examine the effects on attrition of both reenlistment bonuses and reserve pay. They find that the effects of bonuses on attrition are mixed and conclude that "the preponderance of evidence suggests that reenlistment bonuses do not significantly affect attrition" (p. vii). They do report, however, that increases in military pay significantly reduce the rate of attrition. They find that a 10 percent raise in average drill pay reduces attrition by approximately 4.5 percent in the ARNG and 9.5 percent in the USAR. This translates into an attrition pay elasticity—the percentage change in the attrition rate due to a 1 percent change in reserve pay—of about 0.45 for the Guard and 0.95 for the Reserve.

Taking into account the magnitude of the *attrition* rate in the units studied by Marquis and Kirby (about 20 percent annually across both RC), the results also translate into a *retention* pay elasticity of about 0.1 for the Guard and 0.3 for the Reserve. (This is because the retention rate equals one minus the attrition rate, about 80 percent). As noted, retention pay elasticity is a concept slightly different from *reenlistment* pay elasticity—which refers only to the retention decisions at the end of one's term of service. However, bearing in mind that the retention rate in these units (80 percent) was about twice the size of the reenlistment rate (40 percent) in the units studied by Burright et al., the ARNG retention pay elasticity of 0.1 that Marquis and Kirby obtain is similar to the Guard reenlistment pay elasticity of 0.18 estimated by Burright et al. This affords some confidence in the various elasticity estimates.[6]

---

[6]As an example to clarify these relationships, suppose that the pay elasticity for attrition equals 1. If the attrition rate is 20 percent, then a 10 percent increase in pay will

The findings of the foregoing research that attrition is reduced by pay increases or by bonuses requiring retention over a period of time (to receive payment), whereas affiliation bonuses providing full payment more rapidly may not have such effects, suggest that continued payment may be an important component of incentives successful in reducing RC attrition. This suggests use of annual bonuses or increased pay. Note that the numbers also imply that bonuses big enough to bring about large drops in attrition could be on the same order of magnitude as the pay that reservists are now receiving. For example, if it takes a 10 percent increase in drill pay to bring about a 9.5 percent drop in attrition, then we might expect that it would take a bonus exceeding 50 percent of pay to effect a 50 percent attrition reduction. If a 10 percent pay increase reduces attrition by only 4.5 percent, then the required bonus could exceed 110 percent of current pay.

Table 8 shows our estimates of the per-soldier and marginal costs of bonuses required to reduce the percentage of reservists leaving the RC by one-half. The low-cost figure assumes an 0.95 bonus pay elasticity, the high-cost figure an 0.45 bonus pay elasticity. The cost estimates for the average reservist are based on the base case end-strength and years of service distribution by grade. Each bonus includes an increment to compensate the reservist for the reduced chance of promotion that would result if attrition were reduced substantially across the RC. The value of the average reservist's annual pay (including compensation for reduced promotion opportunity) is about $2,850. Since the desired reduction in attrition is 50 percent, the low-cost bonus figure (approximately) equals $2,850 × 0.50/0.95 and the high-cost figure equals $2,850 × 0.50/0.45.

The second row shows the marginal cost of reducing the total number of those who leave by one. Because it is not known who intends to leave, the bonus must be paid to everyone. The ratio of the amounts in the second to the first row indicates that if the attrition rate is halved—from 20 percent to 10 percent—the bonus must be

---

decrease attrition by 10 percent, which equals 2 percentage points. This means that retention will increase by 2 percentage points, which yields a pay elasticity for retention of .25 (i.e., (.02/.80)/.10). If the reenlistment rate were 40 percent and we observed the same increase in retention in response to the pay increase, then the 2-percentage-point increase would equate to a pay elasticity for reenlistment of .5 (i.e., (.02/.40)/.1).

Table 8

Estimated Cost of Bonus to Reduce Attrition by Half

Type of Cost	Cost of Bonus ($)	
	High Elasticity	Low Elasticity
Bonus per soldier	1,500	3,150
Marginal cost	13,200	27,750

paid to the nine of every ten soldiers who remain, only one of whom had intended to leave but is now willing to stay.

In generalizing these results to today's Guard and Reserve, it is important to note that real reserve pay has increased and that real wages for this group have declined since RAND conducted these studies. Thus, an increase in reserve pay now represents a greater fraction of the reservist's total purchasing power than it did in 1982. As a result, reservists' responsiveness to reserve pay could be higher now than it was then.

Emerging results of the current USAR reenlistment bonus program for Contingency Force Pool (CFP) units seem consistent with this notion. (See General Research Corporation, 1995.) The program structure has some basic similarities to the program used in the Reenlistment Bonus Experiment. It is designed to increase six-year reenlistments among soldiers having up to 10 years of service. Money is paid at the time of reenlistment, and again subsequently. In this case, the bonus is $2,500 for a six-year reenlistment, less than today's value of the six-year bonus used in the experiment, and only $500 is paid up front; the remainder is paid annually in installments that increase from $200 to $400 over time.

The results to date for the FY88–94 bonus cohorts suggest that the CFP reenlistment bonus program is having an important effect in reducing attrition. The magnitude of that effect may be two to three times the size of the pay elasticity for attrition reported for the USAR by Marquis and Kirby. More data are needed to be certain, and future comparisons should address the difference in years of service requirements in the various programs. The more senior status of the soldiers eligible for the Reenlistment Bonus Experiment would require a larger bonus to represent the same percentage of pay, and it also could help to explain the lower elasticity, since over time those soldiers wanting to leave the RC would be expected to have done so. Still, the emerging CFP results are encouraging for the types of prospective policies we are evaluating, whose purpose is to reduce personnel turnover throughout the RC, much of which occurs at the junior grade levels.[7]

Reducing Job Turbulence

In addition to reducing reserve attrition, we also recommend reducing job turbulence among reservists as a way to improve DMOSQ rates and job experience levels. Reducing job turbulence is likely to require incentives to encourage individuals to stay in their current occupational specialties. While we are unaware of studies of the cost of incentives to reduce job turbulence per se, we can draw some inferences about the required magnitude of such incentives from the findings of research examining the impact of reenlistment bonuses and pay increases on attrition.

Buddin and Grissmer (1994) report that individual reservists' choices, not Army policy, are primarily responsible for reservists switching MOSs. This implies that the key to decreasing job turbulence is to affect individuals' choices in much the same way that initiatives to reduce attrition do. It seems likely that it would be more difficult to induce someone wanting to leave the reserves entirely not to do so than to induce someone wanting to remain in the RC to remain in his MOS. As a result, we might expect compensation increases sized to reduce attrition to be larger than turbulence-reduction bonuses (and that the former thus could reduce job turbulence to an even greater degree than attrition if made contingent on remaining in the same job).

------

[7]An alternative argument is that studies using the reserve Reenlistment Bonus Experiment data are based on a reserve pool that includes both draft- and non-draft-motivated enlistees, and thus they could overestimate the responsiveness of today's force. Non-draft-motivated reservists tend to have higher retention rates and to reenlist for longer terms. This makes it less likely that a modest reenlistment bonus would have large effects. However, the CFP results to date suggest considerable responsiveness to reenlistment bonuses.

How large do the turbulence-reduction incentives have to be? It is not possible to know the answer with precision absent further research. As suggested earlier in this chapter, however, turbulence might be reduced by a bonus that makes up the pay differential to the reservist's next grade. The number of bonuses offered would depend on the magnitude of the reduction needed.

We examined two variations of this policy. The first involves providing a pool of money for the bonuses tied to the number of job changes and promotions that occur in a one-year period (at each grade level). If one knew exactly which soldiers would be promoted in a given year were they to change jobs, the pool would be used to pay these specific soldiers bonuses exactly equal to the difference in pay to the next-higher grade, given their time in service. Obviously, such omniscience is not possible for either the policymaker or the soldier. Thus, in practice the pool would be used to pay the bonus to a somewhat greater number of soldiers—those with very good chances of promotion were they to change jobs—but at a somewhat reduced value.[8]

This, however, may be a very optimistic approach from a cost standpoint, because it assumes that soldiers understand their true promotion opportunities. An alternative, less-optimistic assumption is that soldiers significantly overestimate their chances of being promoted. If true, it might be necessary to provide bonuses to a much greater number of soldiers in order to reduce job turbulence. In this example, we assume that to reduce turbulence by as much as one-half, the RC might have to pay the bonus to the senior half of the soldiers at every grade level, say according to time in grade. We consider this alternative to provide an upper estimate for turbulence-reduction bonus costs (based on a job change-promotion link).

Table 9 shows the amounts we calculate for bonuses designed to reduce job turbulence by half. Based on the current pay tables, we estimate the average pay increase resulting from promotion to the

---

[8]It may be worth considering some variation in the size of an individual soldier's bonus depending on his true likelihood of being promoted relative to that of the other bonus recipients; however, this would exacerbate within-unit equity concerns (already raised by the payment of the bonus to only some unit members) and would probably introduce debate about the specific criteria to scale down the bonus value

Table 9

Estimated Cost of Bonus to Reduce Job Turbulence by Half

	Cost of Bonus ($)	
Type of Cost	Perfectly Targeted	Pay Top Half
Bonus per soldier	200	200
Marginal cost	200	1,000

next-higher grade (holding years of service constant) at about $200 (top row); as is true for pay, the actual bonus value will vary by grade and years of service. The two columns represent the lower-cost ("omniscient" or perfectly targeted) and higher-cost (pay-the-top-half) cases discussed above. Note that these two different assumptions do not result in different-sized bonuses being paid to an individual soldier.[9] Rather, the assumptions differ in how many soldiers have to be paid that bonus, that is, in our ability to target the bonus. That difference is reflected in the second row, showing the marginal cost of the bonus program per instance of reduced job turbulence (i.e., per soldier persuaded not to change jobs). Note that these turbulence-reduction bonuses are much smaller than the attrition-reduction bonuses in Table 8.

COSTS AND SAVINGS OF ALTERNATIVE TURNOVER-REDUCTION POLICIES

In this section, we combine information we have presented so far in this chapter with information from Chapter Four. Our purpose is to compare the costs of implementing alternative RC personnel turnover-reduction policies with the savings they create from lowering accession and training requirements. We focus on turnover reduction because it offers the greatest potential benefits in raising DMOSQ and job experience levels (see Chapter Four), thereby reducing the prospect of shortfalls in critical personnel during wartime.

---

[9]The bonus value will be reduced in the first case to the extent that targeting those highly likely to be promoted becomes problematic or if its value is adjusted to reflect the soldier's true probability of being promoted if he changes jobs.

We examine the two key policy alternatives to the base case (which, as discussed in Chapter Four, uses recent transition probabilities to project the future RC steady state under current personnel policies). They are a job turbulence reduction of 50 percent and an attrition reduction of 50 percent. For each policy alternative, we estimate the annual savings associated with reduced recruiting and training requirements, as follows. First, we use the readiness enhancement model to estimate the change in the annual accession requirement and in the annual load for each training mode shown in Table 7.[10] The marginal cost figures in Table 7 are then used to calculate the resulting training savings; recruiting savings are calculated as $7,750 per non-prior-service soldier and $2,200 per prior service soldier, as discussed earlier. We also estimate the overall cost of the compensation incentives required by each policy. Consistent with the results reported earlier in Tables 8 and 9, we provide lower and upper estimates for these costs, based on the different assumptions made about the elasticity of compensation on attrition and the ability to target job turbulence reduction bonuses.[11]

---

[10]In essence these are the numbers shown in Table 5. The difference is that for costing purposes we distinguish the AIT course load for MOS retraining from that for initial skill training because of their different costs (see Table 7), and we distinguish non-prior-service accessions from prior-service accessions for similar reasons.

[11]More precisely, the cost is estimated using the readiness enhancement model. At each pay grade level, the model calculates the number of soldiers who must be paid a bonus and, using the dollar values of these bonuses, sums across the pay grades to derive a total cost. In the case of a policy to reduce attrition by one-half (from 20 to 10 percent annually, overall), the bonus is paid to all soldiers retained at each grade level. (This percentage depends on the attrition rate for the given pay grade among non-prior-service and prior-service reservists, and is 90 percent overall.) The bonus value is .50/.95 or .50/.45 times the annual pay of the typical reservist at that grade level.

In the case of a policy to reduce job changes by one-half (from 16 to 8 percent), the number of bonus recipients depends on the targeting assumption. In the perfect targeting case used to derive the lower cost estimate, the bonus is paid to the *additional* soldiers retained in their jobs at each grade level. (This percentage depends on the turbulence rate for the given pay grade among non-prior-service and prior-service reservists, and is 8 percent overall.) The value of the bonus equals the dollar value of promotion to the next-higher pay grade (holding years of service constant). In the high-cost case, the bonus is paid to one-half of *all* soldiers retained in their jobs at each grade level. (This percentage depends on the turbulence rate for the given pay grade among non-prior-service and prior-service reservists, and is 40 percent overall.) In computing cost to the Army, we multiply the cost of the attrition and turbulence reduction bonuses received by each soldier by 1.0765 to account for the Army's FICA obligation.

The results in Table 10 summarize the estimated costs and savings associated with the two alternative policies relative to the base case. We show the costs and savings per 10,000 reservists whom we wish to influence to stay in their jobs or in the RC over the next year—a round number that represents a reasonably sized "building block" for planning the application of incentives to some fraction of the RC.

The turbulence-reduction case shows modest bonus costs. The estimated savings are larger—more than five times larger. This results almost exclusively from training savings. As noted in Chapter Four, there is little effect on accessions; thus, here we see only a very small change in recruiting costs.

The potential savings for attrition reduction are considerably larger; however, the costs are larger still. As noted in Chapter Four, attrition reduction has a considerable effect on accessions; thus, here we see a substantial reduction in recruiting costs, in addition to the potential training savings. But while the total increment in savings relative to turbulence reduction approaches $9 million (per 10,000 reservists), the additional costs are nearly $15 million to $30 million.

Table 10

Estimated Costs and Savings of Alternative Turnover-Reduction Policies (Per 10,000 reservists)

	Cost		Savings	
Policy Yielding Reduction of	Lower Estimate ($M)	Upper Estimate ($M)	Training ($M)	Recruiting ($M)
Turbulence – 50%	0.2	0.8	4.3	0.1
Attrition – 50%	14.6	30.6	7.4	6.0

Chapter Six

BENEFITS, COSTS, AND IMPLICATIONS FOR HIGH-PRIORITY UNITS

We now draw together the results presented in Chapters Four and Five to consider both the improvement in job qualification and experience rates resulting from the implementation of particular policies and the potential costs or savings generated by such policies. We consider these in the light of the shortfalls estimated in Chapter Three. This will lead us to some conclusions about the policies that are needed to address personnel readiness problems in the Army RC, particularly in high-priority units.

COST-BENEFIT OF PROSPECTIVE POLICIES

Table 11 totals the savings from alternative policies and compares them with the costs. In the first two cases, the results are based on Table 10 in the preceding chapter. Recall that those results provided lower and upper estimates of incentive costs. Consequently, the net savings (or cost) figures shown in Table 11 also have lower and upper estimates; the lower estimate equals savings minus maximum cost, and the upper equals savings minus minimum cost.

The job turbulence-reduction figures (first row) simply combine the recruiting and training savings from Table 10 to generate the total estimated savings in Table 11.

The turnover-reduction case (second row) represents equal reductions of 50 percent each in attrition and turbulence. Thus, the potential savings for total turnover reduction combines the effects of the four turbulence and attrition savings cells in Table 10.[1] The incentive costs are those of the attrition case. The logic underlying these costs follows from the ability of properly crafted attrition incentives to reduce job turbulence. This is because the cost analysis suggests that the compensation required to reduce attrition is considerably greater than that required for turbulence reduction. Thus, the attrition-reduction bonus in this alternative should be capable of simultaneously reducing turbulence by 50 percent if made contingent on retention in the same job, obviating the need for a separate turbulence-reduction bonus.[2]

The equal reduction of job turbulence and attrition when examining turnover was useful in assessing readiness effects in Chapter Four. It makes less sense in a cost context, however, given the substantial difference in the cost of the compensation likely to be required to reduce the two types of turnover. Still, the magnitudes and extent of the DMOSQ shortfalls among high-priority units seen in Chapter Three argue for being aggressive in capturing the DMOSQ benefits of reduced attrition. Balancing these two competing motivations, the last row of Table 11 shows a variation on the 50 percent turnover-reduction case in which job turbulence still is reduced by 50 percent but attrition by only 25 percent. The costs are those associated with the attrition reduction. Again, the attrition bonus—half that of the 50 percent attrition-reduction case—would dwarf the turbulence-reduction bonus, suggesting that a separate turbulence bonus is unnecessary for this alternative to reduce turbulence by 50 percent. The savings are approximately one-half of the total attrition-reduction savings shown in Table 10 ($13.4M/2 = $6.7M) plus the full turbulence-reduction savings ($4.4M). As is true for the other cases,

---

[1]The savings are slightly smaller than the simple sum for the four cells; this occurs because some of the retained soldiers change jobs.

[2]It is possible that the elasticity of the attrition-reduction bonus might decline somewhat were payment made contingent on retention in the same job. At the same time, as we noted earlier, (1) there is a wide range in the attrition-reduction bonus estimates and (2) the emerging results from the USAR's current reenlistment bonus program suggest that the bonus's elasticity with respect to attrition may be two to three times greater than that estimated in the earlier research on which the bonus values in Chapter Five are based. Given the extent of uncertainty about the precise bonus value, here we will assume for simplicity that the attrition-reduction bonuses in Chapter Five are capable of jointly reducing attrition and (the post-attrition-reduction rate of) job turbulence by one-half.

Table 11

Net Cost and Savings of Alternative Turnover-Reduction Policies
(Per 10,000 reservists)

Policy Yielding Reduction of	Estimated Cost ($M)		Total Estimated Savings ($M)	Net Savings/(Cost) ($M)	
	Lower Estimate	Upper Estimate		Lower Savings	Upper Savings
Turbulence – 50%	0.2	0.8	4.4	3.6	4.2
Turbulence – 50%, Attrition – 50%	14.6	30.6	17.0	(13.6)	2.4
Turbulence – 50%, Attrition – 25%	7.0	14.6	11.2	(3.4)	4.2

the readiness enhancement model is used to generate the precise cost and savings estimates.

Table 11 shows that to the extent job turbulence can be reduced by paying RC soldiers the dollar value of promotion to the next-higher grade, it will save money. The estimated savings is about $3.6 to $4.2 million per 10,000 reservists. In contrast, cutting attrition by half will probably cost money. The upper estimate for net savings is small and positive ($2.4 million); however, the lower estimate is larger and negative ($13.6 million). Clearly, this is a very wide range; unfortunately, it is driven by the uncertainty surrounding the pay elasticity value. Until additional research can more precisely define this value, it is safest to assume a net cost in the intermediate to high range. Such costs are substantial.

Last, we come to the turnover reduction based on lowering job turbulence by 50 percent and attrition by 25 percent. Given cost considerations, this is probably a more realistic policy goal than one designed to reduce attrition by 50 percent. The results in Table 11 indicate an expected net effect ranging from savings of $4.2 million to costs of about $3.4 million. The midpoint is close to a half-million dollars in net savings per 10,000 soldiers. That is not as much as the net savings from turbulence reduction alone, and it is more uncertain, but it would appear to offer some of the readiness benefits of reduced attrition while avoiding the large expenses of more ambitious attrition-reduction alternatives. It is the readiness benefits of these three policy alternatives to which we now turn.

As shown in Table 12, a bonus paid to reduce job turbulence by 50 percent might be expected to improve the DMOSQ rate by about 9 percent, relative to current policies; depth of job experience would increase by close to 40 percent. An advantage of the job turbulence-reduction bonus is that the policy may not cost anything; on the contrary, we estimate that it would result in a net savings, because the cost of the turbulence-reduction bonus would be outweighed by savings in training costs. (In fact, the bonuses could be five times as big as those we propose and still not result in a net cost to the Army.)

But the turbulence-only approach also has a drawback: As suggested by the results in Chapter Three, many units require improvements in their DMOSQ rates exceeding the 9 percent provided by turbulence reduction if they are to reach their targeted readiness levels. In such cases we need greater leverage. We can accomplish this by tackling attrition reduction along with turbulence reduction. If we provided a bonus to reduce total turnover—both attrition and job turbulence—by 50 percent, we could expect the DMOSQ rate to improve by nearly

Table 12

**Estimated Readiness Benefits and Cost Savings of Alternative
Turnover-Reduction Policies (Per 10,000 reservists)**

Policy Yielding Reduction of	DMOSQ Improvement (%)	Job Experience Improvement (%)	Net Savings/(Cost)	
			Lower Savings ($M)	Upper Savings ($M)
Turbulence – 50%	9.1	38.0	3.6	4.2
Turbulence – 50%, Attrition – 50%	16.6	45.0	(13.6)	2.4
Turbulence – 50%, Attrition – 25%	14.7	56.6	(3.4)	4.2

NOTE: Personnel readiness improvements are relative to the base case. As shown in Table 5, the DMOSQ rates for the first two rows are 68.4 percent and 79.7 percent, respectively. Based on results from the readiness enhancement model, the DMOSQ rate for the last case is 78.4 percent. The job experience improvement is the percent increase in full-time equivalent job years.

17 percent, in relative terms; depth of job experience would increase by 45 percent. These represent large improvements in the units' job qualification levels. Although the potential benefits of such a policy appear to be substantial, our analysis of the applicable costs and savings suggests that it might be very expensive—again, because of the cost of the 50 percent attrition-reduction bonus.

As indicated in previous discussions, a bonus large enough to reduce attrition by 25 percent should still be large enough to reduce job turbulence by 50 percent. The analysis suggests that such a policy could be quite beneficial (third row).   The estimated increase in the DMOSQ rate approaches 15 percent, and depth of job experience increases by 57 percent.[3] The costs for this policy appear to be much more modest than those for a bonus intended to reduce attrition by 50 percent, and may actually result in a net savings.

IMPLICATIONS FOR HIGH-PRIORITY UNITS

Based on these results, what policy might we put into place to address the personnel readiness shortfalls identified for the high-priority units in the SWA scenario, and what would the resulting cost or savings be? For units unable to deploy to the mobilization station at the targeted 85% DMOSQ level:

- We could employ a bonus to reduce job turbulence by half when the required improvement in the DMOSQ rate is below 10 percent. This would provide a 9 percent DMOSQ boost and would most likely cost the Army nothing on net; it would probably even save money.

- We could employ the larger bonus designed to reduce attrition by 25 percent and job turbulence by 50 percent when the improvement needed to reach the targeted DMOSQ level for the unit approaches 10 percent or greater. According to the analysis,

---

[3]Recall that high levels of attrition reduction require increasing the proportion of accessions accounted for by persons in grades E1–E3 to preserve endstrength within grade; most such persons have little or no military experience. This proportion is reduced as the attrition-reduction goal is reduced, explaining the slightly higher level of FEJY in Table 12 for the 25 percent attrition reduction (third row) as compared with the 50 percent reduction (second row).

this would improve the DMOSQ level by about 15 percent and cost the Army less than $3.5 million per 10,000 soldiers, though it might also turn out to be a no-cost or net-savings approach.

These policies would be applied at the unit and MOS level. In this example, only the high-priority units that have DMOSQ shortages for the Southwest Asia scenario would be targeted and, within those units, only specialties whose DMOSQ rates are below 100% of the requirement would receive bonuses. For a specific unit-MOS pair, the type of bonus could be set according to the magnitude of the DMOSQ shortfall of the unit or of the specific MOS within that unit. In the interest of enhancing the readiness of the high-priority units, we favor using the greater of the two shortfalls to determine the bonus type. In the simulated Southwest Asia scenario, this policy resulted in approximately 9,400 soldiers being targeted for turnover reduction bonuses. Of this number, 1,300 were targeted for the job turbulence reduction bonus and the remainder for the 25 percent attrition reduction bonus. The estimated effect of this policy ranges from a net cost of $2.2 million to a net savings of $4.0 million.[4]

UNRESOLVED ISSUES: THE NEED FOR AN EXPERIMENT

Given the foregoing results, it appears that a comprehensive policy to enhance RC personnel readiness and reduce peacetime training demands is feasible. We have discussed the readiness benefits at length above. Other RAND research on the total Army school system is addressing current training demands on RC schools. It has shown that personnel turnover is placing serious strain on that system and creating budgetary pressures, and it suggests that RC schools currently are meeting only 25 percent of the annual training requirement (Winkler, 1995).

While the prospects for improving RC personnel and training readiness appear good, it is very important that turbulence- and attrition-reduction policies be implemented in controlled settings that allow their effects and costs to be systematically evaluated. Given the limits on information currently available about the magnitude of the in-

---

[4]The net cost and savings respectively can be approximated using Table 12, as follows: $(1,300 \times \$3.6M + 8,100 \times -\$3.4M) / 10,000$ and $(1,300 \times \$4.2M + 8,100 \times \$4.2M) / 10,000$.

centives required to carry out such policies and, to a lesser extent, the savings resulting from their implementation, such evaluation is critical in designing the most cost-effective programs.

Among the uncertainties that need to be addressed, three are key: First, what is the actual responsiveness to bonuses of reservists who are thinking of attriting or of changing jobs for reasons other than promotion? The literature gives us some estimates of reservists' responsiveness to attrition-reduction bonuses, but the results are somewhat dated, there is a large range in the elasticity estimates, and the tests were conducted under somewhat different circumstances than would be applicable here. Thus, there is uncertainty. Indeed, emerging results from the USAR's reenlistment bonus program suggest the possibility of considerably greater responsiveness than that suggested by the earlier results.

Similarly, as we noted, for the purpose of the foregoing analysis we assumed that persons changing jobs (and units) do so to increase their promotion prospects. In fact, we believe many do; however, we also believe that many may change jobs for other reasons. One of the unknowns thus involves the size of the bonus that soldiers changing jobs for reasons other than promotion would need to be paid to remain in their jobs and what proportion of the job-changers they represent. Would they want to remain in the RC and thus require a similar (or, perhaps, smaller) bonus than soldiers who change jobs for purposes of promotion? Or would they more closely resemble soldiers thinking of leaving the RC entirely, and thus require a much larger bonus?

A second uncertainty with important cost implications concerns our ability to target job turbulence-reduction bonuses. Under the lower-cost assumption, the pool of money provided for bonuses is constrained to equal the summed pay increases (resulting from promotion to the next-higher grade) for the exact number of additional soldiers we wish to keep in their jobs (relative to the baseline rate of job changes). This amounts to perfect targeting (in terms of the size of the pool). Under the higher cost assumption, we instead assume poor targeting, and plan to pay the 50 percent of the soldiers with the greatest time in grade at each grade level, who may believe they have a reasonable prospect of promotion within the next year. Obviously, the potential cost of the policy varies substantially with our ability to target these bonuses.

Finally, there remains an important issue that we have not yet addressed: the potential impact of the bonuses on the fill rates of the eligible units and MOSs. Until now, we have discussed how these bonuses could lower the rates of job changes and attrition among soldiers assigned to the specific units and MOSs in which they would be applied. But other work—for example, on active duty recruiting—suggests that such bonuses could have market expansion and skill-channeling effects. That is, the bonuses could attract additional, qualified soldiers into the units and specialties offering the bonuses. If that occurred, the bonus policy would produce even greater personnel readiness enhancement for the same cost.

FACTORS ASSOCIATED WITH CROSS-LEVELING

The details provided by the mobilization SMEs together with the analyses in Chapter Two suggest that shortages of DMOSQ personnel were important motivators of the personnel cross-leveling carried out for ODS/S. To examine that relationship directly—that units with lower DMOSQ rates had higher levels of cross-leveling—we employed a regression analysis. The analysis specifically tested for the effects of two factors representing the number of deployable DMOSQ personnel:

- The number of trained personnel whose Primary MOS matched their Duty MOS (computed as a percentage of the unit's required strength)

- The unit's SORTS personnel rating. (Given the DMOSQ variable in the regression, the SORTS rating in this model primarily represents deployability, with higher ratings being better).[1]

The model controlled for several other factors that might influence the extent of cross-leveling in the unit:

- Type of unit (combat or not)

---

[1]AR 220-1 defines the SORTS personnel data rating. The rating reflects assigned vs. required strength (both overall and for senior grades specifically), DMOSQ level among assigned personnel, and availability for deployment with the unit. Since the DMOSQ variable in the regression specification models the percentage of the unit's required positions filled with DMOSQ personnel, in this instance the additional information contributed by the SORTS personnel rating primarily reflects the deployability of these personnel.

- National Guard unit versus Reserve unit
- Whether or not the unit was deployed overseas
- Unit size (assigned personnel).

We used ordinary least squares regression analysis to model the extent of cross-leveling into the activated units—that is, the percentage of personnel deployed with each activated unit who were brought in from other units—as a function of the factors listed above.[2] The cross-leveling regression analysis was carried out for 854 activated units. The results are summarized in Table A.1. The coefficients represent the expected change in (the square root of) the percentage of personnel cross-leveled into a unit given changes in the level of indicated factor.[3] The p-level represents the probability that there was no real relationship between a given unit characteristic and the amount of cross-leveling required for the unit (given the magnitude of the estimated coefficient). Levels below .05 (5 chances in 100) are considered statistically significant.

We found that the unit's DMOSQ level relative to its required strength was a significant predictor of the cross-leveling rate. The lower the DMOSQ level, the greater the cross-leveling. In addition, the readiness level according to the SORTS personnel C-rating was significant—the less deployable the personnel, the greater the cross-leveling. Consistent with their higher C-level requirement, combat units required greater cross-leveling.

When taking these factors into consideration, we found that whether the unit was a Reserve or Guard unit was not a significant factor in altering the rate of cross-leveling. However, the location of deploy-

---

[2]To reduce the error variance resulting from the skew in cross-leveling percentages, we modeled the square root of the percentage cross-leveled.

[3]Dummy variables are coded as 100 or 0 (combat, Reserve, CONUS, SORTS rating missing, fill information missing). DMOSQ rate is a percentage of required strength. The DMOSQ-CONUS interaction variable equals the DMOSQ percentage (less the median of such values, 60 percent) for units deployed in CONUS and equals zero otherwise. The SORTS personnel readiness factor is coded 90, 80, 65, or 50 to reflect comparative personnel readiness C-rating levels. Unit size is coded as the square root of July 1990 strength. Units represent detachments, companies, or battalions (i.e., the lowest meaningful level for deployment purposes and cross-leveling determination identified for the unit in the Unit Identification Code).

Table A.1

Cross-Leveling Regression Summary

Factor	Coefficient	p-level
Combat unit	.0051	.0134
USAR unit	.0010	.3779
Unit deployed in CONUS	−.0041	.0710
DMOSQ rate	−.0279	.0001
SORTS personnel rating	−.0201	.0004
DMOSQ rate, CONUS unit	.0135	.0110
Unit size	−.0456	.0010

NOTE: The following additional coefficients were estimated: intercept, 7.3883; dummy for unknown unit type, .0236; dummy for unknown USAR versus ARNG status, −.0037; dummy for missing readiness information, .0014; dummy for missing DMOSQ rate information, .0001. R-square is .1708.

ment did matter to some extent. Units deployed overseas received marginally more cross-leveling, and the amount of cross-leveling was more responsive to the DMOSQ rate in such units. Finally, unit size also was important. There was a tendency to find greater percentages of cross-leveled personnel in smaller units. This could be due to a reduced ability of one unit member to help cover another's job in small units.

The regression results validate the reports of the subject matter experts we interviewed, and they extend the results from the First Army to the entire Army RC. They support the role of DMOSQ shortages in driving cross-leveling actions during ODS/S, and the potential role of enhanced DMOSQ in reducing the requirement for personnel cross-leveling in future contingencies.

We wish to note that caution is warranted in interpreting the magnitudes of the regression coefficients, which cannot be estimated precisely because of limitations in the available data. Primary among such limitations is missing secondary MOS and training information. We do not know whether a soldier whose Primary MOS was mismatched to his DMOS had a matching secondary MOS. More complete records for the 77th ARCOM suggest that this information would raise a typical unit's DMOSQ rate by only a small amount, about 2 percentage points or less on average; still, for the regression, it is the variation in the secondary MOS match rate across units that is important. Similarly, we do not know exactly how many soldiers still in training in July became MOS qualified before their units deployed. The effect of such instances on the unit's DMOSQ rate is likely to be larger than that of the secondary MOS match rate and more variable across units; available data suggest that the number of DMOSQ personnel in a typical unit would increase by about one-half to two thirds of those in training.

The train-up changes are important in understanding the estimates provided by the model. For example, consider a noncombat ARNG unit of average RC size (about 90 persons) with the average DMOSQ rate (about 63 percent) to be deployed overseas. In July 1990, about 13 percent of required personnel would have been receiving MOS training. By the time of deployment, about 6 to 9 of the 13 percent would have completed training, raising the unit's DMOSQ level to about 70 percent of required strength. The regression indicates that the unit would obtain an additional 18 percent (of its required personnel) through cross-leveling, raising the unit's DMOSQ level to about the 85 percent target level discussed earlier. Alternatively, suppose the unit had a very low DMOSQ rate of only 40 percent. Proportionately, we would expect about 21 percent of required personnel to be in training in July, and about 11 to 14 of the 21 percent to complete training before the unit deployed. This would raise the DMOSQ rate to between 51 and 54 percent of the requirement. The regression indicates that the unit would obtain an additional 24 percent (of its required personnel) through cross-leveling. This would give the unit a DMOSQ rate of roughly 75 to 78 percent, allowing this low-readiness unit to approach the 85 percent DMOSQ target for movement to the mobilization station, and helping to ensure it could meet the C-3 rating required for deployment.

In the preceding paragraph, we provided an example to underscore the importance of accounting for the training pipeline when applying the ODS/S cross-leveling regression results. Given the data limitations described above, we believe the results are more useful for demonstrating the importance of DMOSQ shortages in increasing cross-leveling requirements and for understanding the general patterns of cross-leveling actions undertaken in ODS/S, rather than for predicting precise percentages of cross-leveling for individual units.

THE TRANSPORTATION PROBLEM AND CROSS-LEVELING MODEL

The transportation problem is usually stated in the form popularized by Dantzig (1963). We depart from the usual algebraic notation and use functional notation instead. In functional notation we represent subscripts as function arguments, i.e., "a_i" is written as "a(i)." (This notation corresponds closely with native GAMS notation, the general algebraic modeling system in which the model is currently implemented.)

We modeled the general cross-leveling problem as a variant of the classic "transportation problem" mathematical program. We did this for two reasons: (1) to simplify the problem, so the model developed is easy to validate and verify; and (2) because such models involve dealing with very large data sets—including hundreds of MOSs and units—and there exist special algorithms and techniques for solving large "transportation"-type problems.

THE TRANSPORTATION PROBLEM

In general terms, the transportation problem can be characterized as follows: Given (1) a specified number of manufacturing plants that each have a specified supply of a particular commodity, (2) a specified number of markets for the commodity that each have a specified demand, and (3) the cost per unit shipped between each plant-market pair, what is the optimal shipping strategy that meets the demand at all markets at least cost?

The representation of the transportation problem is:

Indices:

i = plants

j = markets

Given data:

a(i) = supply of commodity at plant i (in cases)

b(j) = demand for commodity at market j (in cases)

c(i, j) = cost per unit shipment between plant i and market j (per case)

Decision Variable:

x(i, j) = number of cases to ship from plant i to market j,

where x(i, j) ≥ 0 for all i, j.

Constraints:

Observe supply limit at plant i:

sum(j, x(i, j)) ≤ a(i) for all i

Satisfy demand at market j:

sum(i, x(i, j)) ≥ b(j) for all j

Objective Function:

Minimize sum((i, j), c(i, j) * x(i, j))

THE CROSS-LEVELING MODEL

The cross-leveling problem can be stated as "given a set of personnel with particular MOSs, DMOSQ status, grades, and unit assignments, what is the best way to reallocate them among MOSs, grades, and units in order to meet a given contingency?" We can map the above formulation of the transportation problem into the cross-leveling problem by thinking of a(i) as corresponding to the number of persons at the start of the simulation with a given MOS, grade, and unit assignment. Similarly, we can think of b(j) as corresponding to the total number of soldiers needed for a particular contingency with a particular MOS, grade, and unit assignment—the "required" strength for the units needed for the contingency, by DMOS-grade pair. The x(i, j) represent the soldiers transferred. In keeping with the transportation problem formulation, all soldiers are "transferred," although since i can equal j, most do not change units. C(i, j) can be thought of as reflecting the relative "costs" or "priorities" of transferring over different specialties, grades, or units. In the implementation below, we map i and j onto *vectors* of characteristics, so the number of indices proliferates, but the basic characteristics of the transportation problem remain constant.

The representation of the problem is:

Indices:

$o = o1 = o2 = MOS$

$g = g1 = g2 = grades$

$u = u1 = u2 = units$

Given data:

$a(o1, g1, u1)$ = supply of personnel with MOS o1, grade g1, in unit u1

$b(o2, g2, u2)$ = demand for personnel with MOS o2, grade g2, in unit u2

$c(o1, g1, u1, o2, g2, u2)$ = cost/priority for transfers of various types

Decision Variable:

$x(o1, g1, u1, o2, g2, u2)$ = number of personnel with MOS o1, grade g1, in unit u1, assigned to MOS o2, grade g2, in unit u2, where $x(o1, g1, u1, o2, g2, u2) \geq 0$ for all o1, g1, u1, o2, g2, u2.

Constraints:

Observe supply limit:
$sum((o2, g2, u2), x(o1, g1, u1, o2, g2, u2)) \leq a(o1, g1, u1)$
for all o1, g1, u1.

Satisfy demand:
$sum((o1, g1, u1), x(o1, g1, u1, o2, g2, u2)) \geq b(o2, g2, u2)$
for all o2, g2, u2.

Objective Function:
Minimize $sum((o1, g1, u1, o2, g2, u2),$
$c(o1, g1, u1, o2, g2, u2) * x(o1, g1, u1, o2, g2, u2)).$

The cost function is simple, but it can work effectively to reflect relative priorities of cross-leveling over grades, across units, or within MOSs. Very high costs can be used to restrict the allowable transitions (i.e., reflecting the one-up, two-down rule, or allowable units to draw from, etc.); however, in the GAMS implementation, we have used dummy variables to indicate allowable transitions.

In the model runs used for the analysis, the cost/priority of the various transitions is generated by a fairly simple algorithm:

$$c(o1, g1, u1, o2, g2, u2) = 0 \quad \text{to start}$$

	add 1	if there is a grade mismatch (i.e., $g1 \neq g2$)
	add 10	if there is a unit mismatch (i.e., $u1 \neq u2$)
	add 100	if there is a MOS mismatch (i.e., $o1 \neq o2$)

After optimizing the allocation of soldiers within their unit to DMOS for which they are qualified by occupation and grade, this gives first priority to filling shortfalls by moving people between grades, second priority to moving people between units, and third priority to giving people special training to take care of a shortfall.

This results in the following costs/priorities:

$$
\begin{aligned}
c(o1, g1, u1, o2, g2, u2) \quad &= 0 && \text{if } o1 = o2 \text{ and } u1 = u2 \text{ and } g1 = g2 \\
&= 1 && \text{if } g1 \neq g2 \\
&= 10 && \text{if } u1 \neq u2 \\
&= 11 && \text{if } g1 \neq g2 \text{ and } u1 \neq u2 \\
&= 100 && \text{if } o1 \neq o2 \\
&= 101 && \text{if } o1 \neq o2 \text{ and } g1 \neq g2 \\
&= 110 && \text{if } o1 \neq o2 \text{ and } u1 \neq u2 \\
&= 111 && \text{if } o1 \neq o2 \text{ and } u1 \neq u2 \text{ and } g1 \neq g2
\end{aligned}
$$

The formulation uses only three equations, and avoids the multiplication of transition functions of an "inventory" specification of the problem. Implicitly, any move or any combination of moves is allowed. The model can be extended easily by lengthening the vector of characteristics associated with supply or demand, for example, adding additional units and occupations to the model. Dummy variables can be added to reflect constraints on allowable transitions (e.g., the "one-up, two-down" rule for allowable grade substitutions, MOS-grade restrictions on permissible cross-training or deployment schedule-based constraints on receiving cross-trainees, deployment schedule-based constraints on cross-leveling actions, etc.).

Stating Demand in Terms of Unit DMOSQ Level

In the above model, requirements are given independently for each occupation and grade within a unit. However, the cross-leveling policy usually is carried out in a "unitwide" form, in terms of a DMOSQ level for the whole unit. This discrepancy can be addressed by adding the desired DMOSQ level for each unit and adding an additional "demand" equation that computes the overall rating, unit by unit. Also, in the above model, requirements were made absolute and no shortfall was allowed. In the actual model, units are allowed to fall short of the requirement in particular occupations and grades as long as the overall unit DMOSQ target is met. However, they are not allowed to receive cross-leveled soldiers in excess of the requirement in particular occupations and grades for the purpose of meeting the unit DMOSQ target.

Training and Time

Notions of time and the possibility of training to fill needed occupations can be treated by extending the vector of characteristics to include a time index, and by allowing transitions that require training only if sufficient time is available. That is, availabilities and requirements are given by:

$a(o1, g1, u1, t1) =$ supply of personnel with MOS o1, grade g1, in unit u1 available at time t1

$b(o2, g2, u2, t2) =$ demand for personnel with MOS o2, grade g2, in unit u2 required at time t2

Training feasibility is given by the following matrix of allowable transitions:

train (o1, t1, o2, t2)   = 1   if there is sufficient time between t1
                               and t2 to train an o1 into an o2

                         = 0   otherwise

In all other respects the problem is identical with the ones specified above.

Revised Model

A bar ("|") has been placed in the first column where there have been changes from the first statement of the model above.

Indices:
   o = o1 = o2 = MOS

   g = g1 = g2 = grades

   u = u1 = u2 = units

Given data:

   a(o1, g1, u1) = supply of personnel with MOS o1, grade g1, in unit u1

   b(o2, g2, u2) = demand for personnel with MOS o2, grade g2, in unit u2

   c(o1, g1, u1, o2, g2, u2) = cost/priority for transfers of various types

   dmosq(u2) = desired DMOSQ rating of each unit

   cshort1(o2, g2, u2) = cost of falling short of o2, g2, u2

   cshort2(u2) = cost of falling short of u2 requirement

Decision Variables:

   x(o1, g1, u1, o2, g2, u2) = number of personnel with MOS o1, grade g1, in unit u1 assigned to MOS o2, grade g2, in unit u2, where x(o1, g1, u1, o2, g2, u2) ≥ 0 for all o1, g1, u1, o2, g2, u2.

short1(o2, g2, u2) = shortfall in requirement for o2, g2, u2.

short2(u2) = shortfall in requirement for u2.

Constraints:

Permit shortfall within each occupation and grade, relative to unit DMOSQ requirement:

sum((o1, g1, u1), x(o1, g1, u1, o2, g2, u2))

+ short1(o2, g2, u2) ≥

dmosq(u2) * b(o2, g2, u2) for all o2, g2, u2.

Permit shortfall from minimum DMOSQ requirement for each unit:

sum((o1, g1, u1, o2, g2), x(o1, g1, u1, o2, g2, u2))

+ short2(u2) ≥

dmosq(u2) * sum(o2, g2), b(o2, g2, u2)) for all u2.

Objective Function:

Minimize sum((o1, g1, u1, o2, g2, u2),

c(o1, g1, u1, o2, g2, u2) * x(o1, g1, u1, o2, g2, u2))

+ cshort1(o2, g2, u2) * short1(o2, g2, u2)

+ cshort2(u2) * short2(u2).

RESULTS FOR THE NORTHEAST ASIA, SOUTHWEST ASIA, AND TWO-MRC SCENARIOS

Below, Table B.1 shows the percentage of units unable to deploy at their targeted DMOSQ levels in the three different scenarios we modeled. The results for the Northeast Asia and Southwest Asia scenarios are similar. For the two-MRC scenario, the percentage of units having problems under the "all units" cross-leveling plan increases, because more units must be deployed and, therefore, there are fewer nondeploying units to cross-level from. The percentage of problem units under the "units deploying after 30 days" plan declines somewhat, because the delay in the onset of the second MRC

Table B.1

**Percentage of Units Unable to Deploy at Targeted
DMOSQ Level for Three Scenarios**

	Type of Cross-Leveling Planned	
Scenario	All Units	Units Deployed After 30 Days
Northeast Asia	0.6	22.5
Southwest Asia	2.1	25.6
Two MRCs	11.6	20.1

NOTE: Percentages represent shortfall as a percentage of all units
to be deployed. The number of units required for the three modeled scenarios are 485, 566, and 1,024, respectively.

means that a smaller *percentage* of the deploying units leave within
the first 30 days. Of course, the total *number* of problem units is
larger.

Table B.2 shows a full example of the detailed personnel realignment
and cross-leveling actions for one sample unit. The "MOS" and
"GRADE" columns show the occupation and grade respectively. The
"START" column shows how many DMOSQ people are available
within the unit, including those currently holding other jobs or
whose regular training will be completed by the mobilization date.
The next eight columns show the ways in which people can be re-
aligned or assigned to the unit: The "NO CHANGE" column shows
the number of people that remain in the same MOS and grade;
"CHANGE GRADE" shows the people that moved into a particular
MOS and grade by either moving up one grade or down as much as
two grades while remaining in the same unit and MOS; "CHANGE
UNIT" shows the number of people that transferred into the unit, re-
taining the same MOS and grade, and so on. The "FINAL TOTAL"
column shows the number of people in a given MOS and grade after
all the realignment and cross-leveling is done; it is the same as the
total of the preceding eight columns (from "NO CHANGE" to
"CHANGE MOS GRADE UNIT"). The "DMOSQ% • REQUIREMENT"
column shows the product of the targeted DMOSQ percentage for
the unit and the number of soldiers required in wartime for the MOS-
grade pair, whereas the "100% REQUIREMENT" column shows the

total wartime requirement for the same MOS-grade pair. The final column shows the percentage of the total requirement attained within each MOS-grade pair. (Of course, DMOSQ is properly looked at as a unitwide figure, but the breakdown by MOS-grade pair is illuminating and, as noted, can be used to help choose the appropriate personnel readiness enhancement policy for the specific MOS and unit.)

In this table one can easily see grade movements within the unit in the column "CHANGE GRADE"—one example is MOS 55B, where an E5 is assigned to a slot normally requiring an E3. One can also observe where there are "excess" people in the unit who were retrained to fill a needed MOS in "CHANGE MOS." By analogy, the same information is provided for soldiers cross-leveling into the unit.

Figures B.1 and B.2 show the effects of the turnover-reduction policies discussed in Chapter Six together with those of cross-leveling for the Southwest Asia scenario in which cross-leveling is planned for units deploying after C+30. Figure B.1 shows the effect of the turnover-reduction policies on the readiness of units with DMOSQ shortfalls, before any cross-leveling takes place. It indicates that such policies could reduce the number of units with the most serious personnel readiness shortfalls—DMOSQ levels below C-3—by one-third (from 91 to 58) and nearly triple of the units at C-1 to C-2 (from 19 to 53). Figure B.2 adds cross-leveling to the equation for units deploying after C+30 and shows the final DMOSQ percentage distribution of the entire force deploying for the scenario. Most of the shortfalls are eliminated by the combination of turnover-reduction and cross-leveling.

Table B.2

Detailed Breakdown of Personnel Realignment and Cross-Leveling Actions for One Unit

MOS	GRADE	START	NO CHANGE	CHANGE GRADE	CHANGE UNIT	CHANGE GRADE UNIT	CHANGE MOS	CHANGE MOS GRADE	CHANGE MOS UNIT	CHANGE MOS GRADE UNIT	FINAL TOTAL	DMOSQ%•REQIREMENT	100% REQIREMENT	%OF REQIREMENT
00Z	E9	1	1								1	0.85	1	100
31U	E1–3			2							2	1.70	2	100
31U	E4				2						2	1.70	2	100
31U	E5				2						2	1.70	2	100
31U	E7				1						1	0.85	1	100
52D	E4				1						1	0.85	1	100
52F	E4					1					1	0.85	1	100
54B	E5	1	1								1	0.85	1	100
54B	E6	1	1								1	0.85	1	100
55B	E1–3		1	5							6	5.10	6	100
55B	E4			4							4	3.40	4	100
55B	E5	3	2								2	1.70	2	100
55B	E6	1	1								1	0.85	1	100
63B	E1–3			2							2	1.70	2	100
63B	E4	1	1	2							3	2.55	3	100
63B	E5		1	1							2	1.70	2	100
63B	E6	2	1								1	0.85	1	100

Table B.2—Continued

MOS	GRADE	START	NO CHANGE	CHANGE GRADE	CHANGE UNIT	CHANGE GRADE UNIT	CHANGE MOS	CHANGE MOS GRADE	CHANGE MOS UNIT	CHANGE MOS GRADE UNIT	FINAL TOTAL	DMOSQ% • REQUIREMENT	100% REQUIREMENT	% OF REQUIREMENT
63B	E7	1	1								1	0.85	1	100
63J	E4				1						1	0.85	1	100
63S	E1–3	2	1								1	0.85	1	100
63S	E4			1	1						2	1.70	2	100
63S	E5	1	1								1	0.85	1	100
67R	E1–3				1		5				6	5.10	6	100
67R	E4	10	7								7	5.95	7	100
67R	E5	17	17	2							19	16.15	19	100
67R	E6	7	7								7	6.80	8	87
67R	E7	6	6								6	6.80	8	75
67T	E1–3						1				1	0.85	1	100
67T	E4	1	1								1	0.85	1	100
67T	E5	3	3	1							4	3.40	4	100
67T	E6	3	2								2	1.70	2	100
67V	E1–3	1	1	1							2	1.70	2	100
67V	E4	7	7								7	5.95	7	100
67V	E5	9	5								5	4.25	5	100
67V	E6	7	5								5	4.25	5	100
67Z	E8	4	4								4	3.40	4	100
68B	E4	1	1								1	0.85	1	100
68B	E5						1				1	0.85	1	100

Table B.2—Continued

MOS	GRADE	START	NO CHANGE	CHANGE GRADE	CHANGE UNIT	CHANGE GRADE UNIT	CHANGE MOS	CHANGE MOS GRADE	CHANGE MOS UNIT	CHANGE MOS GRADE UNIT	FINAL TOTAL	DMOSQ %• REQlREMENT	100% REQIREMENT	% OF REQIREMENT
68D	E4	3	1								1	0.85	1	100
68D	E5	2	1								1	0.85	1	100
68F	E1–3						1			1	2	1.70	2	100
68F	E4	1	1								1	0.85	1	100
68F	E5	1	1								1	0.85	1	100
68G	E1–3	1	1								1	0.85	1	100
68G	E4			1							1	0.85	1	100
68G	E5	2	1								1	0.85	1	100
68H	E5											0.85	1	
68J	E1–3						1			3	4	3.40	4	100
68J	E4	1	1			2				3	6	5.10	6	100
68J	E5	1	1								1	3.40	4	25
68J	E6	1	1								1	2.55	3	33
68J	E7											0.85	1	
68K	E7	1	1								1	0.85	1	100
68N	E1–3	1	1		1						2	1.70	2	100
68N	E4			2							2	1.70	2	100
68N	E5	3	2								2	1.70	2	100
68N	E6	1	1								1	0.85	1	100
71D	E5			1							1	0.85	1	100
75B	E4	1												

Table B.2—Continued

MOS	GRADE	START	NO CHANGE	CHANGE GRADE	CHANGE UNIT	CHANGE GRADE UNIT	CHANGE MOS	CHANGE MOS GRADE	CHANGE MOS UNIT	CHANGE MOS GRADE UNIT	FINAL TOTAL	DMOSQ%. REQUIREMENT	100% REQUIREMENT	% OF REQUIREMENT
75B	E5	1	1	1	1						3	2.55	3	100
75B	E6	1	1								1	0.85	1	100
75Z	E7	1	1								1	0.85	1	100
77F	E1–3	1	1		4						5	4.25	5	100
77F	E4	3	3		3						6	5.10	6	100
77F	E5	7	7								7	5.95	7	100
77F	E6	2	1								1	0.85	1	100
77F	E7	1	1								1	0.85	1	100
91B	E1–3	1	1								1	0.85	1	100
91B	E4	1	1								1	0.85	1	100
91B	E5	2	1								1	0.85	1	100
92A	E1–3				3						3	2.55	3	100
92A	E4				1						1	0.85	1	100
92A	E5				2						2	1.70	2	100
92Y	E4				5						5	4.25	5	100
92Y	E6				1						1	0.85	1	100
92Y	E7				1						1	0.85	1	100
93B	E4	1												
93B	E5	2	2	1							3	2.55	3	100
93B	E6	2												
93P	E4	2	2		2						4	3.40	4	100

Table B.2—Continued

MOS	GRADE	START	NO CHANGE	CHANGE GRADE	CHANGE UNIT	CHANGE GRADE UNIT	CHANGE MOS	CHANGE MOS GRADE	CHANGE MOS UNIT	CHANGE MOS GRADE UNIT	FINAL TOTAL	DMOSQ% • REQIREMENT	100% REQIREMENT	% OF REQIREMENT
93P	E5	3	3	1	6						10	8.50	10	100
93P	E6	1												
93P	E7				1						1	0.85	1	100
93P	E8	1	1			1					2	1.70	2	100
94B	E1–3			1	1						2	1.70	2	100
94B	E4	1	1	2							3	2.55	3	100
94B	E5	4	1								1	0.85	1	100
94B	E6				1						1	0.85	1	100
94B	E7	1	1								1	0.85	1	100
96B	E4	1	1		1						2	1.70	2	100
96B	E5	2	2								2	1.70	2	100

NOTE: This example is for the Southwest Asia scenario.

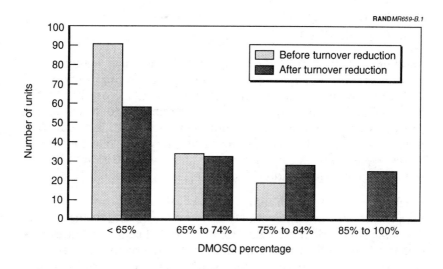

Figure B.1—DMOSQ Distribution of Early-Deploying Units with Shortfalls
Before and After Turnover Reduction
(SWA Scenario, No Cross-Leveling in First 30 Days)

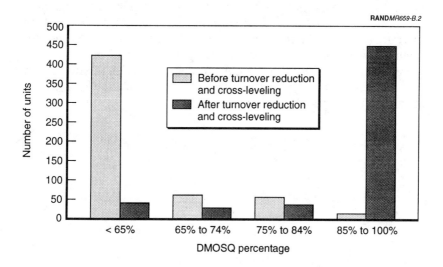

Figure B.2—DMOSQ Distribution of All Deploying Units Before and After
Turnover Reduction and Cross-Leveling
(SWA Scenario, No Cross-Leveling in First 30 Days)

ADDITIONAL CONSIDERATIONS IN ADDRESSING PERSONNEL READINESS SHORTFALLS

Another consideration in the application of personnel readiness enhancement policies could be to fine tune the attrition-reduction level and related bonus size depending on the particular features of a targeted MOS. Specialties with high attrition rates and/or long training programs could be targeted for more ambitious attrition-reduction policies—because the savings would be greater—while those with below-average values could be targeted for lower attrition reduction or for turbulence reduction alone.

Finally, there may be additional methods the Army can use to help obtain the substantial benefits of lowered attrition at reduced cost. The analyses conducted by Burright et al. and Marquis and Kirby (see Chapter Four) show that attrition varies by demographic characteristics such as age, marital status, gender, civilian occupation, and others, and that the impact of these characteristics on the probability that an individual leaves the reserve can be as large as the impact of a monetary retention incentive such as a bonus or pay increase. Thus, just as the DoD learned that recruiting high school graduates into the active force substantially reduced attrition, it appears that focusing on recruits with particular demographic characteristics could reduce reserve attrition. The relative cost-effectiveness of such targeted recruiting programs in lowering attrition would depend on the cost of recruiting individuals with the desired characteristics relative to the cost of recruiting other individuals and the size of the reduction in attrition elicited by selective recruiting.

THE READINESS ENHANCEMENT MODEL: TRANSITION PROBABILITIES

In the readiness enhancement model described in Chapter Four, transition probabilities govern how personnel change states during an annual cycle, including the simultaneous change of two or three states and the possibility of remaining unchanged. There is a probability for separation and a probability for each of the eight possible combinations of three other events: promotion or not; stay in same job ("stable") or not; DMOSQ or not ("NQ"). For each of the personnel inventory categories—described in Chapter Four—these nine probabilities sum to one.

We reproduce below the tables of transition probabilities that we used in the base case. Multiplying the transition probabilities by the start-of-year inventory yields personnel flows between categories. Table C.1 shows the steady-state base case RC enlisted inventory. Table C.2 shows the steady-state transition probabilities applicable to the base case for the postdrawdown force. Table C.3 shows the multipliers applied to the raw FY93 transition probabilities (not shown) to produce the steady-state transition rates in Table C.2.

Table C.1

Steady-State Base Case Inventory

Non-Prior Service	Number	Prior Service	Number
E1–3 New NQ	50,939	E1–3 New NQ	1,018
E1–3 New DMOSQ	5,305	E1–3 New DMOSQ	1,578
E1–3 Old NQ	18,718	E1–3 Old NQ	556
E1–3 Old DMOSQ	41,580	E1–3 Old DMOSQ	567
E4 New NQ	0	E4 New NQ	4,676
E4 New DMOSQ	2,615	E4 New DMOSQ	14,354
E4 Old NQ	14,375	E4 Old NQ	8,045
E4 Old DMOSQ	65,946	E4 Old DMOSQ	23,168
E5 New NQ	1,084	E5 New NQ	2,927
E5 New DMOSQ	1,563	E5 New DMOSQ	5,005
E5 Old NQ	8,581	E5 Old NQ	10,520
E5 Old DMOSQ	35,365	E5 Old DMOSQ	35,319
E6 New NQ	286	E6 New NQ	580
E6 New DMOSQ	314	E6 New DMOSQ	962
E6 Old NQ	4,723	E6 Old NQ	7,240
E6 Old DMOSQ	19,380	E6 Old DMOSQ	27,443
E7–9 New NQ	0	E7–9 New NQ	0
E7–9 New DMOSQ	54	E7–9 New DMOSQ	397
E7–9 Old NQ	3,500	E7–9 Old NQ	7,766
E7–9 Old DMOSQ	11,287	E7–9 Old DMOSQ	22,264
TOTAL	285,615	TOTAL	174,385

Table C.2

Steady State Base Case Transition Probabilities

	Grade	Leave	Stay Within Grade				One Grade Promotion			
			Stable		Move		Stable		Move	
			NQ	DMOSQ	NQ	DMOSQ	NQ	DMOSQ	NQ	DMOSQ
NPS	E1–3 New NQ	0.2331	0.1443	0.4284	0.0878	0.0400	0.0046	0.0494	0.0061	0.0062
	E1–3 New Q	0.3920	0.0092	0.1742	0.0586	0.0136	0.0021	0.2654	0.0551	0.0297
	E1–3 Old NQ	0.3406	0.1543	0.1633	0.0792	0.0464	0.0363	0.1239	0.0240	0.0320
	E1–3 Old Q	0.2822	0.0023	0.2957	0.0497	0.0120	0.0014	0.3008	0.0351	0.0206
	E4 New NQ	0.3079	0.1614	0.2652	0.1211	0.0692	0.0074	0.0461	0.0074	0.0144
	E4 New Q	0.3407	0.0082	0.4232	0.1021	0.0395	0.0010	0.0666	0.0068	0.0121
	E4 Old NQ	0.2558	0.2538	0.1751	0.1225	0.1148	0.0134	0.0361	0.0102	0.0183
	E4 Old Q	0.2156	0.0026	0.5722	0.0759	0.0327	0.0010	0.0841	0.0086	0.0074
	E5 New NQ	0.2220	0.2038	0.2652	0.1654	0.0874	0.0070	0.0304	0.0070	0.0117
	E5 New Q	0.2765	0.0090	0.4681	0.1138	0.0469	0.0023	0.0587	0.0124	0.0124
	E5 Old NQ	0.1587	0.3176	0.1877	0.1372	0.1299	0.0117	0.0299	0.0113	0.0160
	E5 Old Q	0.1507	0.0036	0.6582	0.0840	0.0327	0.0007	0.0591	0.0059	0.0051
	E6 New NQ	0.1961	0.3009	0.2563	0.1393	0.0836	0.0060	0.0060	0.0060	0.0060
	E6 New Q	0.2236	0.0041	0.5245	0.1301	0.0610	0.0000	0.0393	0.0131	0.0044
	E6 Old NQ	0.1258	0.3515	0.1823	0.1425	0.1368	0.0085	0.0260	0.0101	0.0165
	E6 Old Q	0.1007	0.0036	0.7238	0.0870	0.0384	0.0009	0.0365	0.0042	0.0048
	E7–9 New NQ	0.2595	0.2645	0.2380	0.1058	0.1322				
	E7–9 New Q	0.2492	0.0215	0.5685	0.1180	0.0429				
	E7–9 Old NQ	0.1002	0.4049	0.1448	0.2170	0.1331				
	E7–9 Old Q	0.0801	0.0026	0.7871	0.0967	0.0334				

Table C.2—Continued

Grade	Leave	Stay Within Grade				One Grade Promotion			
		Stable		Move		Stable		Move	
		NQ	DMOSQ	NQ	DMOSQ	NQ	DMOSQ	NQ	DMOSQ
PS									
E1-3 New NQ	0.2931	0.1695	0.0809	0.1063	0.0231	0.0706	0.1516	0.0477	0.0572
E1-3 New Q	0.3677	0.0013	0.1486	0.0584	0.0152	0.0018	0.3138	0.0652	0.0281
E1-3 Old NQ	0.3972	0.1811	0.0660	0.1008	0.0366	0.0595	0.0949	0.0310	0.0329
E1-3 Old Q	0.4582	0.0007	0.2473	0.0422	0.0089	0.0015	0.2024	0.0249	0.0139
E4 New NQ	0.3254	0.2103	0.1838	0.1299	0.0749	0.0085	0.0433	0.0077	0.0161
E4 New Q	0.3344	0.0036	0.4433	0.0867	0.0348	0.0009	0.0790	0.0079	0.0093
E4 Old NQ	0.3039	0.2508	0.1500	0.1232	0.0791	0.0145	0.0475	0.0116	0.0193
E4 Old Q	0.2428	0.0034	0.4965	0.0767	0.0347	0.0012	0.1195	0.0134	0.0118
E5 New NQ	0.2534	0.2420	0.1914	0.1588	0.0875	0.0075	0.0346	0.0084	0.0164
E5 New Q	0.2639	0.0039	0.4755	0.0990	0.0427	0.0007	0.0942	0.0105	0.0096
E5 Old NQ	0.1718	0.3071	0.1615	0.1517	0.1161	0.0148	0.0425	0.0139	0.0205
E5 Old Q	0.1388	0.0037	0.6360	0.0884	0.0398	0.0007	0.0752	0.0083	0.0091
E6 New NQ	0.2516	0.2573	0.1988	0.1476	0.0976	0.0078	0.0183	0.0092	0.0118
E6 New Q	0.2581	0.0080	0.4702	0.1048	0.0609	0.0032	0.0721	0.0140	0.0086
E6 Old NQ	0.1317	0.3466	0.1591	0.1591	0.1315	0.0114	0.0235	0.0143	0.0228
E6 Old Q	0.1020	0.0034	0.6854	0.0880	0.0561	0.0006	0.0513	0.0057	0.0074
E7-9 New NQ	0.2099	0.2584	0.1640	0.2336	0.1342				
E7-9 New Q	0.1517	0.0088	0.6919	0.0749	0.0727				
E7-9 Old NQ	0.1066	0.4009	0.1160	0.2383	0.1382				
E7-9 Old Q	0.0870	0.0054	0.7628	0.1017	0.0432				

Table C.3

Adjustments to Raw FY93 Transition Probabilities

Grade	Promotion	Attrition	Turbulence
Non-prior service			
E1–3	0.80	0.90	1.00
E4	1.26	0.98	1.00
E5	1.12	0.90	1.00
E6	1.07	0.92	1.00
E7–9		0.92	1.00
Prior service			
E1–3	0.80	0.90	1.00
E4	1.26	0.98	1.00
E5	1.12	0.90	1.00
E6	1.07	0.92	1.00
E7–9		0.92	1.00

TRAINING SAVINGS

This appendix documents fully the methodology for calculating dollar savings from reductions in training load. We present both the method we applied in the model, which calculates incremental savings from changing the training load, and an alternate method based on net incremental savings.

INCREMENTAL SAVINGS METHODOLOGY

Using a "bottom-up" method, we first compiled the incremental cost of Advanced Individual Training (AIT) in Active Component schools at the level of individual MOSs. Of course, the cost of training different MOSs varies considerably depending on the length and resource demands of the course. To consolidate to a single set of factors, we calculated a weighted average course cost for each reserve component, where the weights reflected expected requirements for level one (initial) training in each MOS. We also calculated costs for Reserve Component Training Institutions (RCTIs) and for Basic Training. The incremental cost of each of these various training modes represents the potential savings when that particular training load is reduced.

Below, we document the estimation process just described.

AIT Resource Requirements

Training requires military and civilian manpower, funding for students attending courses, ammunition, the provision of training supplies and materials, and the provision of installation support. Changes in training load have a potential effect on the demand for each of these resources. Below, we list the training cost components categorized by appropriation.

Direct Training Cost

- Military Personnel
  - School staff pay and allowances
  - Student pay and allowances
- Operations and Maintenance (O&M)
  - Civilian pay
  - Other O&M (e.g., Temporary Duty (TDY); Petroleum, Oil, and Lubricants (POL); repair parts)
- Procurement
  - Ammunition for training

Indirect Training Cost (i.e., Base Operations)

- Military Personnel
  - Base staff pay and allowances
- Operations and Maintenance
  - Civilian pay
  - Other O&M (e.g., TDY; POL; repair parts)

While there are additional components in the total cost of training (e.g., cost of training development), the above categories cover all the incremental costs of training.

For most cost elements, we used TRADOC published factors to calculate the incremental cost per course of changing the training load for AIT. The Deputy Chief of Staff for Resource Management (DCSRM) of TRADOC generally uses factors published yearly in the *TRADOC Resource Factor Handbook* to estimate marginal funding adjustments in TRADOC schools. Although grouped by school and installation, the factors for school manpower, O&M, and Base Operating Support (BOS) cover every course taught at TRADOC installations, where RC soldiers receive AIT. Student pay and allowances, one of two cost elements not covered by the TRADOC handbook, was computed by applying pay and allowance factors from the Force and Organization Cost Estimating System (FORCES) and the *Reserve Forces Almanac* to the length of individual courses. The cost of ammunition for each course was obtained from the Army Manpower Cost System (AMCOS) data file, maintained by the Army's Cost and Economic Analysis Center (CEAC).

The TRADOC method uses the following input factors, one set for each TRADOC school and installation, to calculate the incremental cost of training:

School Cost Factors:

a. Total Operations and Maintenance (O&M) cost per student year

b. Percent of Operations and Maintenance (O&M) representing civilian pay

c. Staff required per 100 student years

d. Percent of required staff that are authorized

e. Percent of authorized staff that are officers

f. Percent of authorized staff that are warrant officers

g. Percent of authorized staff that are enlisted

h. Annual pay and allowances, officers

i. Annual pay and allowances, warrant officers

j. Annual pay and allowances, enlisted

Installation Support Cost Factors:

k. Total Operations and Maintenance (O&M) cost per student year

l. Percent of Operations and Maintenance (O&M) representing civilian pay

m. Staff required per 100 student years

n. Percent of required staff that are authorized

o. Percent of authorized staff that are officers

p. Percent of authorized that are warrant officers

q. Percent of authorized staff that are enlisted

r. Annual pay and allowances, officers

s. Annual pay and allowances, warrant officers

t. Annual pay and allowances, enlisted

Computing Weighted-Average AIT Costs

Because we did not specify training load reductions by MOS, we had to consolidate widely varying individual course costs into an average cost per course, accounting for large differences in course demand and length. We approximated the distribution of demand for training in different MOSs by the number of E1 through E4 positions authorized in FY93.[1] For example, because MOS 11B represented 10.8 percent of E1 through E4 authorizations in the Guard (the most common occupation), we used that percentage factor as a weight to calculate the cost of an average course in the Guard. In contrast, the cost of the 11B course received a weight of only 2.8 percent in the calculation of average course costs in the USAR, because there are fewer soldiers with that MOS in that component. Using the authorization information to rank order the MOSs by size, we selected the set of largest MOSs—those accounting for 90 percent of total authorizations—to use to compute the weighted average cost in each component.

The weighted costs per student undergoing MOS reclassification training in an active AIT school for an "average" course appear in Table D.1 below. (Student pay and allowances show figures that are

---

[1] In effect, this assumes that the average cost per course in the postdrawdown RC will be about the same as in FY93 (adjusted to constant dollars). This assumption may be conservative; given the greater postdrawdown concentration of support forces in the RC and the greater average length of the related MOS courses relative to combat MOS courses, the actual cost may increase, and with it the potential savings from reductions in personnel turnover.

Table D.1

Average Reclassification AIT Course Cost Per Student

Type of Cost	Guard ($)	Reserve ($)	All RC ($)
Direct cost			
School staff pay and allowances	2,068	1,480	1,878
Student pay and allowances	7,329	7,091	7,242
Civilian pay at school	259	204	241
Other Operations and Maintenance (O&M)	113	79	102
Ammunition for training	303	148	253
Indirect cost			
Installation staff pay and allowances	103	100	102
Installation support O&M	389	362	380
Total cost	10,564	9,464	10,198

halfway between the cost of an E4 and an E5.)[2] An "All RC" figure is obtained by weighting the Guard and Reserve figures by the relative number of E1 to E4 authorizations. The slightly higher figures in the Guard column imply that the distribution of MOSs in that component are somewhat more expensive to train.

The Cost of Other Training Modes

Not all MOS retraining takes place in active schools. Most retraining of RC soldiers occurs in Reserve Component Training Institutions (RCTIs). Generating appropriate assumptions for the reserve schools requires recognition of their differences from active schools. First, RC schools teach from a program of instruction (POI) that is configured for RC training periods, specifically, 38 days annually. Completing a course generally stretches out over an entire year (sometimes more) in that training mode, with about two-thirds of the course occurring during annual training and the other third during weekend training. As a result, the RC-configured POI involves fewer training

---

[2]Costs for initial AIT following Basic Training would be less because of lower student pay and allowances. Costs for this and other forms of training are taken up below.

days than the AC POI; we estimate (using the weighted average method described above) that the average RC course is about one-third the length of an average AC course.

Second, RCTIs have different staffing structures than active schools. Because RCTIs are designed to provide training near the reservist's home station, they tend to be multifunctional rather than specialized, and smaller but more numerous than active schools. In addition, because they represent a dispersed school system, they are not able to achieve the same economies of scale as the larger active schools. The three major types of RCTI providing MOS training are U.S. Army Reserve Forces schools (USARF), State Military Academies (SMA) in the Guard, and more specialized Reserve Training Sites-Maintenance (RTSM). Our factor for school staff pay and allowances in RCTI represented a weighted average of the costs for the three major school types based on their relative student loads.

In addition to the cost of reclassification training, we also examined the cost of initial entry training for soldiers with no prior military service. Initial entry training involves Basic Training and AIT. Concerning the former, we estimated the cost based on the methodology described in the *TRADOC Resource Factor Handbook* (p. 22 of the August 1994 version).

In Table D.2 we summarize the cost of four training types: Basic Training; AIT training when taken for the first time, following Basic Training; AIT training meant to reclassify a trained soldier into a different MOS; and RCTI training meant to reclassify a trained soldier. These costs also represent the factors we use in Chapters Four and Five to estimate the cost savings resulting from alternative readiness enhancement policies.

The major difference among the first three columns is in student pay and allowances, accounted for by the differing grades of the trainees: an E1 for Basic Training, an E2 for AIT after BT, and between an E4 and E5 for AIT for MOS reclassification. The major difference between reclassification in an AIT course versus an RCTI course is in the shorter length (about one-third as long) of the latter. However, the cost of an RCTI reclassification is about half as great (rather than a third), because school staff pay and allowances are higher for RCTI courses than for AIT courses. This reflects the greater fixed cost of a

Table D.2

Costs (Potential Per-Soldier Savings) by Type of Training

Type of Cost	Basic Training ($)	AIT after BT ($)	AIT for Reclass. ($)	RCTI ($)
Direct cost				
School staff pay and allowances	1,408	1,878	1,878	2,283
Student pay and allowances	3,747	4,773	7,242	2,087
Civilian pay at school	0	241	241	217
Other Operations and Maintenance (O&M)	25	102	102	46
Ammunition for training	486	253	253	115
Indirect cost				
Installation staff pay and allowances	102	102	102	33
Installation support O&M	380	380	380	123
Total cost	6,148	7,729	10,198	4,903

dispersed system of schools, one that must give up the economies of scale available to the larger active schools in order to bring training closer to the student.

NET INCREMENTAL SAVINGS

The incremental cost savings from reducing the training load does not necessarily represent the *net dollar effect*, because some of the savings can be realized in different ways. In particular, since variances in training load may not affect endstrength decisions, savings in military pay and allowances may be realized in terms of freed military manpower rather than dollars. For example, if training reductions led to 100 fewer required authorizations at schools, 100 spaces would be freed up for other missions. These spaces might be assigned to TOE units, in which cases readiness enhancements rather than dollar savings accrue.

Further, because of current budgetary pressures and, among soldiers with substantial employment in the civilian labor force, possible civilian employer restrictions on taking the additional 2–3 weeks required to receive both MOS training and AT, today's RC soldiers often attend school in an "alternate AT" or "alternate IDT" status (meaning they attend school instead of, rather than in addition to, attending normal unit training exercises). In such cases, much of the potential dollar savings would not be realized if training were reduced, because similar expenses would be incurred at the unit training site if the student were not attending school. Instead of occurring as dollar savings, benefits would be realized in terms of increases in unit readiness, as more soldiers attend training with their unit. Recent surveys of RC commands suggest that few students attend both school and unit training within the same month, and only about 7 percent of those attending RCTIs attend two AT sessions.

More generally, the problem involves the lack of a meaningful Trainees, Transients, Holdees, and Students (TTHS) account for the RC. Such accounting is critical to the AC. Soldiers are not assigned to operational units or counted against unit strength while they are being trained. This maintains the DMOSQ rates and readiness levels within units, and enhances collective training. We believe that the same approach would be highly desirable for the future RC personnel system; however, we do not assume it in our models because it seems unlikely to be established in the near term. Instead, we assume that policies to enhance RC personnel readiness include overcoming current obstacles to providing IDT and AT to (re)trainees as part of the strategy. Thus, in computing potential savings in future training costs resulting from RC personnel readiness enhancement policies, we report incremental savings.

In this appendix, however, we disaggregate the potential savings into its component parts given current conditions. The net dollar savings from a MOS reclassification training load reduction for AIT are calculated in Table D.3. In the first column we show the incremental savings from AIT training load reductions by type of cost. The remainder of the table breaks down these savings as follows: net dollar savings actually realized in the budget, costs shifted to unit training, and the value of freed manpower. Of the $10,200 in savings, we estimate that $1,952 will be realized in terms of freed manpower spaces, and that $2,254 will be offsets to unit training. The latter figure derives from the assumption that trainees will not attend AT or weekend drills with their unit while attending the AIT course (of approximately 9 weeks).

Savings in school staff pay and allowances can also be expressed in terms of the number of personnel. In Table D.4, we estimate the number of personnel that would be freed up from a reduction in training load of a thousand student slots in AIT courses.

Table D.3

Net Dollar Savings Per Student for AIT (Reclassification)

Type of Cost	Total Savings ($)	Unit Training Offsets ($)	Value of Freed Manpower ($)	Net Dollar Savings ($)
School staff pay and allowances	1,878	0	1,878	0
Student pay and allowances	7,242	2,021	0	5,221
Civilian pay at school	241	0	0	241
Other training Operations and Maintenance (O&M)	102	28	0	74
Ammunition for training	253	71	0	182
Installation staff pay and allowances	102	28	74	0
Installation support O&M	380	106	0	274
Total cost	10,198	2,254	1,952	5,992

Table D.4

Total Personnel Savings Per Thousand AIT Students

Type of Personnel	Number
Training personnel:	
Officers	3.3
Warrant Officers	0.5
Enlisted	39.7
Installation support personnel:	
Officers	0.2
Enlisted	2.2

In Table D.5, we estimate the net dollar savings from a training load reduction in RC schools. Notice that while potential savings in RCTI are about 50 percent of those in AIT courses in AC schools, the current net dollar savings would be less than 15 percent ($734 compared to $5,992). This occurs because today RCTI courses are configured to the existing RC training schedule (instead of being given in a continuous block), making more opportunity for unit training offsets. Thus, reducing the need for RCTI courses simply displaces the bulk of the student effort to other RC activities, instead of eliminating the time and saving the money.

In Table D.6, we summarize the factors used to calculate net cost savings for the four training types we addressed: Basic Training, AIT training when taken immediately following Basic Training, AIT training meant to reclassify a trained soldier into a different MOS, and RC school training meant to reclassify a trained soldier. Note that the two training situations not discussed above—BT and AIT after BT—involve no dollar offsets to unit training, since it is assumed unit training is not an alternative to Basic Training and initial AIT.

Table D.5

Net Dollar Savings Per Student in RCTI

Type of Cost	Total Savings ($)	Unit Training Offsets ($)	Value of Freed Manpower ($)	Net Dollar Savings ($)
School staff pay and allowances	2,283	0	1,895	388
Student pay and allowances	2,087	1,973	0	113
Civilian pay at school	217	0	0	217
Other training Operations and Maintenance (O&M)	46	44	0	3
Ammunition for training	115	108	0	6
Installation staff pay and allowances	33	31	2	0
Installation support O&M	123	116	0	7
Total cost	4,903	2,273	1,896	734

Table D.6

**Decomposition of Per-Soldier Savings from Reduced
Training, by Type of Training**

Type of Cost	Basic Training ($)	AIT after BT ($)	AIT for Reclass. ($)	RCTI ($)
Net dollar savings	4,638	5,749	5,992	734
Dollar offsets in unit training	0	0	2,254	2,273
Dollar value of freed manpower	1,510	1,980	1,952	1,896
Total cost (incremental savings)	6,148	7,729	10,198	4,903

BIBLIOGRAPHY

Asch, Beth J., and James N. Dertouzos, *Educational Benefits Versus Bonuses: A Comparison of Recruiting Options.* Santa Monica, CA: RAND, MR-302-OSD, 1994.

Buddin, Richard J., and Carole E. Roan, *Assessment of Combined Active/Reserve Recruiting Programs.* Santa Monica, CA: RAND, MR-504-A, 1994.

Buddin, Richard J., and David W. Grissmer, *Skill Qualification and Turbulence in the Army National Guard and Army Reserve.* Santa Monica, CA: RAND, MR-289-RA, 1994.

Burright, Burke K., David W. Grissmer, and Zahava D. Doering, *A Model of Reenlistment Decisions of Army National Guardsmen.* Santa Monica, CA: RAND, R-2866-MRAL, October 1982.

Dantzig, George B., *Linear Programming and Extensions.* Princeton, NJ: Princeton University Press, 1963.

Fernandez, R. L., *Enlistment Effects and Policy Implications of the Educational Assistance Test Program.* Santa Monica, CA: RAND, R-2935-MRAL, September 1982.

General Research Corporation, Data on retention effects of USAR bonus programs. Vienna, VA, 1995.

Goldberg, Larry, *Estimates of the Marginal Costs of Selected Supply Factors Based on Recent Enlistment Supply Analyses.* Economic Research Laboratory, March 1985.

Grissmer, David W., Richard J. Buddin, and Sheila Nataraj Kirby, *Improving Reserve Compensation: A Review of Current Compensation and Related Personnel and Training-Readiness Issues.* Santa Monica, CA: RAND, R-3707-FMP/RA, September 1989.

Grissmer, David W., Zahava D. Doering, and Jane Sachar, *The Design, Administration, and Evaluation of the 1978 Selected Reserve Reenlistment Bonus Test.* Santa Monica, CA: RAND, R-2865-MRAL, July 1982.

Hosek, James R., and C. E. Peterson, *Reenlistment Bonuses and Retention Behavior.* Santa Monica, CA: RAND, R-3199-MIL, 1985.

Kirby, Sheila Nataraj, and David W. Grissmer, *Reassessing Enlisted Reserve Attrition: A Total Force Perspective.* Santa Monica, CA: RAND, N-3521-RA, 1993.

Marquis, M. Susan, and Sheila Nataraj Kirby, *Reserve Accessions Among Individuals with Prior Military Service: Supply and Skill Match.* Santa Monica, CA: RAND, R-3892-RA, October 1989a.

Marquis, M. Susan, and Sheila Nataraj Kirby, *Economic Factors in Reserve Attrition: Prior Service Individuals in the Army National Guard and Army Reserve.* Santa Monica, CA: RAND, R-3686-1-RA, March 1989b.

National Defense Research Institute (NDRI), *Assessing the Structure and Mix of Future Active and Reserve Forces: Final Report to the Secretary of Defense.* Santa Monica, CA: RAND, MR-140-1-OSD, 1992.

Office of the Inspector General, *Special Assessment, National Guard Brigades' Mobilization.* Department of the Army, June 1991.

Polich, J. Michael, James M. Dertouzos, and James Press, *The Enlistment Bonus Experiment.* Santa Monica, CA: RAND, R-3353-FMP, April 1986.

Reserve Forces Almanac, Uniformed Services Almanac Inc., Falls Church, VA, 1994.

Shukiar, Herbert J., *Readiness Enhancement Model: Inventory Projection Model of the Army's Reserve Components.* Santa Monica, CA: RAND, MR-659/1-A (forthcoming).

Sortor, Ronald E., *Army Active/Reserve Mix: Force Planning for Major Regional Contingencies.* Santa Monica, CA: RAND, MR-545-A, 1995.

Sortor, Ronald E., Thomas F. Lippiatt, and J. Michael Polich, *Planning Reserve Mobilization: Inferences From Operation Desert Shield.* Santa Monica, CA: RAND, MR-123-A, 1993.

Sortor, Ronald E., Thomas F. Lippiatt, J. Michael Polich, and James C. Crowley, *Training Readiness in the Army Reserve Components.* Santa Monica, CA: RAND, MR-474-A, 1994.

Tan, Hong W., *Non-Prior Service Reserve Enlistments: Supply Estimates and Forecasts.* Santa Monica, CA: RAND, R-3786-FMP/RA, 1991.

TRADOC Resource Factor Handbook, published by Deputy Chief of Staff for Resource Management (DCSRM), Program Analysis and Evaluation (PA&E), TRADOC, August 1994.

U.S. Army, *Unit Status Reporting,* AR-220-1.

U.S. General Accounting Office, National Security and International Affairs Division, *National Guard Peacetime Training Did Not Adequately Prepare Combat Brigades for Gulf War.* Washington, D.C.: GAO, GAO/NSIAD-91-263, 1991.

U.S. Government, *Army National Guard Combat Readiness Reform Act of 1992 (Title 11).* P.L. 102-484, Sec. 1111–1137, 1992.

Winkler, John D., *Restructuring the Total Army School System.* Santa Monica, CA: RAND, DB-153-A, 1995.